Fair Trade and Organic Agriculture

A Winning Combination?

Fair Trade and Organic Agriculture

A Winning Combination?

Edited by

Priyanka Parvathi

Ulrike Grote

Hermann Waibel

CABI

CABI is a trading name of CAB International

CABI
Nosworthy Way
Wallingford
Oxfordshire OX10 8DE
UK

Tel: +44 (0)1491 832111
Fax: +44 (0)1491 833508
E-mail: info@cabi.org
Website: www.cabi.org

CABI
745 Atlantic Avenue
8th Floor
Boston, MA 02111
USA

Tel: +1 (617)682-9015
E-mail: cabi-nao@cabi.org

A catalogue record for this book is available from the British Library, London, UK.

Library of Congress Cataloging-in-Publication Data

Names: Parvathi, Priyanka, editor.
Title: Fair trade and organic agriculture : a winning combination? / edited by Priyanka Parvathi, Ulrike Grote & Hermann Waibel.
Description: Boston, MA : CABI, 2017. | Includes bibliographical references and index.
Identifiers: LCCN 2017029487 (print) | LCCN 2017031465 (ebook) | ISBN 9781786394316 (pdf) | ISBN 9781786393074 (ePub) | ISBN 9781786393050 (hardback : alk. paper)
Subjects: LCSH: Organic farming. | Anti-globalization movement.
Classification: LCC S605.5 (ebook) | LCC S605.5 .F35 2017 (print) | DDC 631.5/84--dc23
LC record available at https://lccn.loc.gov/2017029487

ISBN-13: 9781786393050 (hbk)
9781786394316 (ePDF)
9781786393074 (ePub)

Commissioning editor: David Hemming
Associate editor: Alexandra Lainsbury
Production editor: Tim Kapp

Typeset by SPi, Pondicherry, India
Printed and bound in the UK by Bell & Bain Ltd, Glasgow, G46 7UQ

Contents

Contributors

Arnab K. Basu, Charles H. Dyson School of Applied Economics and Management, Cornell University, Ithaca, New York, USA
Lee Byers, Coffee and Tea, Global Product Management, Fairtrade International, Bonn, Germany
Rene T. Capote-Fuentes, Coffee, Global Product Management, Fairtrade International, Bonn, Germany
Ricardo Fort, GRADE Group for the Analysis of Development, Lima, Peru
Ulrike Grote, Institute for Environmental Economics and World Trade, Leibniz University Hannover, Germany
Casey Goldvale, Georgetown University, Washington, DC, USA
Ulrich Hamm, Department of Agricultural and Food Marketing, University of Kassel, Germany
Robert Hicks, Department of Economics, College of William and Mary, Williamsburg, Virginia, USA
Pradyot Ranjan Jena, School of Management, National Institute of Technology, Karnataka, India
Linda Kleemann, Research Area Poverty Reduction, Equity and Development, Kiel Institute for the World Economy.
Katharina Krumbiegel, University of Goettingen, Department of Agricultural Economics and Rural Development, Germany
Julia Lernoud, Department of Extension, Training and Communication, Research Institute of Organic Agriculture (FiBL), Frick, Switzerland
Priyanka Parvathi, Institute of Development and Agricultural Economics, Leibniz University Hannover, Germany
Yuhui Qiao, College of Resources and Environmental Sciences, China Agricultural University, Beijing, China
Ruerd Ruben, Wageningen Economic Research, Wageningen University and Research Centre, The Netherlands
Rosa Schleenbecker, Department of Agricultural and Food Marketing, University of Kassel, Germany
Till Stellmacher, Center for Development Research (ZEF), University of Bonn, Germany
Romane Viennet, Cornell University, Ithaca, New York, USA and Sciences Po Paris, France
Hermann Waibel, Institute of Development and Agricultural Economics, Leibniz University Hannover, Germany
Helga Willer, Department of Extension, Training and Communication, Research Institute of Organic Agriculture (FiBL), Frick, Switzerland

Meike Wollni, University of Goettingen, Department of Agricultural Economics and Rural Development, Germany
Katrin Zander, Thünen Institute of Market Analysis, Federal Research Institute for Rural Areas, Forestry and Fisheries, Braunschweig, Germany

Acknowledgements

We acknowledge the financial support of *Wege in die Forschung II* (Project funding for young scientists, Leibniz University Hannover) that enabled us to conduct and organize an international workshop in connection with this book. We thank all the contributors, authors and co-authors of the book for submitting their ideas and chapters as well as for playing a major role in the shaping of this book. In addition to submitting their own chapters; many authors and co-authors also thoroughly reviewed other chapters and we gratefully acknowledge this input. We tremendously appreciate their time and effort invested in the preparation of this book.

As many of the case studies presented in this book are based on extensive field research, we would like to acknowledge all inputs from consumers and producers who participated in the data collection survey and made these studies possible.

At CABI we would like to thank Rachel Cutts for her vital role in the inception of this book and to Dr David Hemming for his continued support and patience in the making of this book. His inputs are valued, appreciated and acknowledged. We also express our thanks to Emma McCann and Alexandra Lainsbury for their support from the administrative side in the development of this book.

1 An Overview of Organic Agriculture and Fair Trade Systems

Priyanka Parvathi* and Hermann Waibel
Institute of Development and Agricultural Economics, Leibniz University Hannover, Germany

1.1 Introduction

Since the Brundtland Commission coined the term 'sustainable development' in its report *Our Common Future* (Brundtland Commission, 1987), this approach has increasingly gained global prominence. The concept relating to agriculture and rural development has been at the heart of many discussions among supporters and sceptics of sustainability. In this context, eco-friendly and ethical aspects of production like organic agriculture and fair trade have been discussed. Also global awareness concerning economic development, social equity and environmental protection has grown considerably.

In international agricultural debates, certification systems like Fair Trade and organic farming are considered as serving niche markets. Fair Trade certification is used as a unique selling proposition in markets like coffee, banana, cocoa, mango and traditional handicrafts. Organic certification is more centred on high-value markets like cotton, tea, coffee and spices. In recent years, organic fruit and vegetables have also captured consumer interest in the developed nations. Though extensive agricultural debates on these subjects are lacking, both these certification systems provide a possibility for agriculture to diversify into non-traditional methods of production and agricultural marketing.

The idea of Fair Trade has its roots in world trade. Nevertheless, it has opened new agricultural market prospects. The inherent strength and advantage of a Fair Trade certification for agricultural produce are in providing a rural, poor and remote smallholder farmer with access to global markets. Fair Trade has the potential to provide development opportunities and better living conditions for poor farmers in developing countries.

Organic agriculture, on the other hand, is a technical innovation that is believed to be environmentally friendly and ecologically sustainable. However, it is viewed as an infeasible strategy for global agriculture, due to food security aspects. The arguments against organic agriculture in meeting global food supply demands are predominantly concerned with low yields (Rigby and Cáceres, 2001). Nevertheless, organic produce has a niche market in the developed world for its food safety and quality.

The global market for both these innovations has grown rapidly, especially over the past decade. However, the share of these products at retail level remains small, though many studies have documented willingness to pay for organic

*Corresponding author; e-mail: parvathi@ifgb.uni-hannover.de

and Fair Trade products. The global sales of Fair Trade goods were €7.3 billion in 2015 (Willer and Lernoud, 2016). The global market size of organic produce has increased threefold in the past 10 years and was valued at €75.7 billion in 2015 (Willer and Lernoud, 2016). Though these products are assumed to appeal to ethically and environmentally conscious consumers, who are considered a minority, these sales figures indicate that their number has been increasing in recent years. The expanding markets and growing sales indicate sustained and increasing consumer demand for these certified commodities. Thus there is merit in assessing whether it is technically and economically feasible to meet these growing consumer demands in relation to standards of food safety, quality and ethics, especially concerning smallholders and marginal producers. Perhaps it is time to study these certification systems as emerging areas of agricultural research.

Both these certification systems critique conventional agriculture and seek to create an eco-friendly agronomy and smallholder producer development (Raynolds, 2000). While Fair Trade as a movement was implemented from developing countries, organic agriculture evolved in the developed nations. Each certification system caters to different aspects of agriculture: organic is production specific; whereas Fair Trade relates to marketing of farm produce.

The literature deliberates on organic farming and Fair Trade regimes. Some prominent examples include Browne *et al.* (2000), Rice (2001), Raynolds (2004), Calo and Wise (2005) and Bacon (2008). Though combining these innovations helps to reduce farmers' livelihood vulnerability (Bacon, 2005), it is also contended that such certification systems alone do not provide clear advantages to smallholder farmers (Valkila, 2009). So far, research has treated the two systems separately and little is known about whether adopting both systems in combination might be more beneficial for small-scale farmers in developing countries. Also it needs to be explored whether consumers are willing to pay more for a product that is both organic and Fair Trade certified.

The question remains: can adopting both these certification systems together contribute to the agricultural and socio-economic development of smallholder producers in developing countries? This book contributes to the debate and fills the gap in the literature by pursuing a twofold aim: (i) to assess the welfare impacts of either system or of both systems in combination on smallholders in developing countries; and (ii) to examine the willingness of consumers in the developed world to pay for such certified produce.

1.2 Certification Bodies

Organic agriculture and Fair Trade systems are governed by two separate certification bodies. While the International Federation of Organic Agriculture Movements (IFOAM) is the umbrella organization for organic agriculture, Fairtrade Labelling Organizations International (FLO) sets Fair Trade certification standards.

1.2.1 IFOAM

IFOAM has its roots in the international congress on organic agriculture held in Versailles, France, on 5 November 1972. The five founding members of IFOAM included: the French farmer organization *Nature et Progrès*; Rodal Press in the USA; the Soil Association in the UK; the Soil Association of South Africa; and the Swedish Biodynamic Association (Paull, 2010). IFOAM has grown rapidly and today it has 800 affiliates across more than 100 countries. IFOAM defines organic agriculture as follows.

> Organic Agriculture is a production system that sustains the health of ***soils, ecosystems*** and ***people***. It ***relies on ecological processes, biodiversity and cycles adapted to local conditions***, rather than the use of inputs with adverse effects. Organic Agriculture combines ***tradition, innovation*** and ***science*** to benefit the shared environment and promote ***fair relationships*** and a good ***quality of life*** for all involved.

Its standards are primarily based on the following four principles (IFOAM, 2006; Geier, 2007):

- ***Principle of health:*** Organic Agriculture should sustain and enhance the health of soil, plant, animal, human and planet as one and indivisible.
- ***Principle of ecology:*** Organic Agriculture should be based on living ecological systems and cycles, work with them, emulate them and help sustain them.

- ***Principle of fairness:*** Organic Agriculture should build on relationships that ensure fairness with regard to the common environment and life opportunities.
- ***Principle of care:*** Organic Agriculture should be managed in a precautionary and responsible manner to protect the health and well-being of current and future generations and the environment.

IFOAM launched the International Organic Accreditation Service (IOAS) in 1997 to accredit third-party certification bodies to provide organic certification. It assesses the competence of these organizations to issue organic certification. If the accreditation is achieved, then the control institution is approved to confer organic certification to farmers. It is also agreed that the certification body will certify farmers based on the rules of IFOAM standards. Broadly, a third-party certification body needs to study in detail the processes that are allowed and not allowed under organic farming principles (including storage) as specified by IFOAM. When a farmer applies to a third-party certification body for organic certification, the steps listed below are broadly followed.

- If the farmer is converting from conventional farming to organic, normally a 3-year conversion period is estimated. Conversion periods could be longer, depending on the intensity of chemical use on the farmland in the past.
- Farmers have to provide extensive documentation on farm history, production practices and results of farm soil and water analysis.
- A production plan must also be submitted that details the source of seeds, material inputs used, harvesting methods employed and storage facilities.
- The certifying body then sends their trained representative to inspect the farm, examines the records submitted and submits an evaluation report.
- To receive organic certification the farmer has to pay the inspection fees and, upon a positive evaluation, also the certification fees.
- Furthermore, certified farmers must continue to keep day-to-day records of all farming activities and these records need to be available any time for inspection.
- The certifying body needs to make surprise and short-notice inspections and undertake specific tests, such as soil analysis, from time to time. Costs of such inspections are borne by the farmer.

Apart from implementing private organic standards, IFOAM is currently engaged with governmental and intergovernmental standard-setting processes such as the European Union (EU) regulations on organic agriculture. It is also engaged with international organizations such as the Food and Agriculture Organization of the United Nations (FAO), United Nations Conference on Trade and Development (UNCTAD), World Trade Organization (WTO), United Nations Environment Program (UNEP), the International Labour Organization (ILO) and the Organization for Economic Cooperation and Development (OECD), among others (Luttikholt, 2007; Paull, 2010). However, it needs to be noted that accreditation through IFOAM does not equally mean compliance with national organic standards of all importing countries. It is possible that certain countries may ask for additional requirements to be complied with before accepting an imported product as organic.

IFOAM is now usually known as 'IFOAM – Organics International'.

1.2.2 FLO

The first Fair Trade label, 'Max Havelaar', was launched in 1988 by Solidaridad, a Dutch development agency, to transport Fair Trade coffee from Mexico to supermarkets in The Netherlands. This initiative was replicated in many countries, leading to prominent Fair Trade labels like 'TransFair' and 'Fair Trade Mark'. The initial Fair Trade certifications were restricted to coffee. In 1997 Fairtrade International was launched in Bonn, Germany, to unite all Fair Trade organizations under one umbrella. This also enabled the development and harmonization of all global Fair Trade certification standards. In 2004 Fairtrade International split into FLO and FLO-CERT. FLO sets Fairtrade standards and also provides support to producers. FLO-CERT is the independent certifier of Fairtrade products; it conducts inspections and awards certification to producer organizations and plantations and it also audits traders.

The size of an association certified as Fair Trade can range from small farmer cooperatives

Box 1.1. Fair Trade glossary

The *Fair Trade Glossary*, published jointly in 2011 by the World Fair Trade Organization, Fairtrade International and FLO-CERT, clarifies the difference between 'Fair Trade' (sometimes abbreviated to FT) and 'Fairtrade' (no abbreviation) as follows.

Fair Trade

The term Fair Trade defines a trading partnership, based on dialogue, transparency and respect, that seeks greater equity in international trade. It contributes to sustainable development by offering better trading conditions to, and securing the rights of, marginalized producers and workers – especially in developing countries.

Fairtrade

Fairtrade refers to all or any part of the activities of FLO eV, FLO-CERT, Fairtrade producer networks, national Fairtrade organizations and Fairtrade marketing organizations. Fairtrade is used to denote the product certification system operated by Fairtrade International (FLO).

Fairtrade International / Fairtrade Labelling Organizations International eV (FLO)

Fairtrade International (FLO) is a multi-stakeholder, non-profit organization focusing on the empowerment of producers and workers in developing countries through trade. FLO provides leadership, tools and services needed to connect producers and consumers, promote fairer trading conditions and work towards sustainable livelihoods.

Fairtrade Labelling Organizations International eV is the legally registered name for 'Fairtrade International'.

FLO-CERT GmbH (FLO-CERT)

FLO-CERT GmbH is the independent Fairtrade certification body offering Fairtrade certification services to clients in more than 70 countries. FLO-CERT, a private limited company, evaluates Fairtrade certification applications, verifies compliance with the Fairtrade Standards during audits, and decides whether Fairtrade certification can be granted or not. Certificates are issued as proof of successful Fairtrade certification, and as authorization to trade Fairtrade products. FLO-CERT holds an ISO65 accreditation and follows best practice in all of its certification operations.

World Fair Trade Organization (WFTO)

The World Fair Trade Organization is a global network of Fair Trade organizations and WFTO associates representing the supply chain from producer to retailer. WFTO certification includes, for example, handcraft producers, not simply agricultural commodities.

to large plantations with hundreds of workers as well as large cooperatives with thousands of farmers as members. If organizations apply for Fair Trade certification with FLO-CERT, it performs producer audits by physically visiting and interviewing a minimum number of farmers from the applicant organization. This audit will last for several days, depending on the size of the applicant as well as the different number of products that need Fair Trade certification. The auditor's physical report is then submitted to FLO-CERT for approval or to follow up on any non-conformity. FLO-CERT evaluates the report and awards Fair Trade certification, provided that all non-conformities (if any) are corrected after a follow-up visit. Thereafter FLO-CERT physically visits certified organizations and conducts announced audits at least twice in three years, as well as carrying out surprise audits. The annual cost of certification depends on the total number of farmers or workers in the applicant organization.

Traders applying to sell Fair Trade certified products temporarily receive immediate permission to sell on receipt of the application. This temporary permission lasts for a maximum of 6 months, within which time the first Fair Trade audit needs to be conducted. The auditor checks if the Fair Trade premium and minimum prices are paid in accordance with Fair Trade standards.

As with producer audits, traders are also subjected to an announced audit at least twice in three years as well as surprise audits.

1.3 Organic Agriculture and Fair Trade: a Comparison

As explained above, organic agriculture and Fair Trade have separate certification bodies. They differ with respect to scope, consumer motivation, certification, microfinance institutions and price mechanisms, as depicted in Table 1.1 (Parvathi and Waibel, 2013).

Organic agriculture is concerned with farm production processes, whereas Fair Trade mainly focuses on marketing agricultural and other products (e.g. traditional handicrafts) that also meet certain social standards. Fair Trade standards differentiate between smallholder and plantation producers. Smallholders are mostly required to organize themselves into cooperatives or producer organizations to apply for Fair Trade certification. Large plantations can directly apply for Fair Trade certification provided that they meet the required Fair Trade standards for hired labour. On the contrary, organic farming is scale neutral; individual farmers need to apply independently to get their land certified as organic.

In developing countries, producers under Fair Trade can get up to 60% advance credit from their buyers if requested, as noted by Dragusanu *et al.* (2014), enabling them to avoid falling into debt traps to meet rising input costs for materials. However, Fair Trade goes against free market price mechanisms and defines two sets of prices. It sets a minimum price, which is expected at least to cover the cost of production, as well as a price premium. The cooperative or producer organization that sells under a Fair Trade scheme can sell its produce at the market price, or at the minimum set Fair Trade price if it is higher than the market price. This protects the smallholders from market price shocks. The price premium from Fair Trade is expected to be invested by the cooperative in social and rural development initiatives such as improving infrastructure, water systems, building schools and maintaining empowerment programmes for women. Fair Trade also helps the cooperative to establish long-term relationships with the buyers of its produce and to benefit from the presence of an assured buyer from whom it could also receive advance payments when required. Certified organic produce can be sold at premium prices above the market prices of its counterpart conventional produce. Many studies have argued that this premium pricing of organic products is essential to counter the low yields that are generally associated with eco-friendly farming systems (e.g. Tzouvelekas *et al.*, 2001; Lansink *et al.*, 2002; Kumbhakar *et al.*, 2009).

The niche markets for organic and Fair Trade, especially in the developed world, have witnessed continued growth. Consumers in the developed world prefer organic to conventional produce due to its food safety, quality and eco-friendly production aspects. The majority of Fair Trade and organic produce is from the developing continents of Asia, Africa and Latin America, as depicted in Table 1.2 (Parvathi and Waibel, 2013). Fair Trade buyers in industrialized nations are driven by ethical and social concerns like poverty alleviation. Thereby these systems also capture the interdependence between the developed and the developing nations in promoting eco-friendly and sustainable development. Moreover, both these certification systems provide a possibility for agriculture to diversify into non-traditional methods of production and agricultural marketing.

Table 1.1. Organic and fair trade comparison (modified from Parvathi and Waibel, 2013).

Criteria	Organic	Fair Trade
Scope of standard	Production and product-focused	Marketing and labour conditions at work
Target group	Small- and large-scale farmers	Mostly small-scale farmers; can be extended to large-scale farmers if they meet additional labour conditions
Consumer motivation	Environment and health	Poverty reduction
Microfinance institutions	No specific credit programmes	Possibility of advance payments
Price	Market price	Market price or floor price, whichever is higher

Table 1.2. Producers and buyers of organic and Fair Trade (Parvathi and Waibel, 2013).

	Organic	Fair Trade
Major producers	Asia, Australia and Latin America (Willer and Kilcher, 2011)	Asia, Africa and Latin America (Krier, 2008)
Major buyers	Europe and North America (Willer and Kilcher, 2011)	Europe, North America, Japan, Australia and New Zealand (Krier, 2008)

In recent years, the markets for products that are both organic and Fair Trade have increased and continue to grow rapidly. This raises questions about whether it would benefit smallholder producers in developing countries to adopt these two systems in combination, whereby they could ensure that their products are both ethically friendly and environmentally viable. However, this would mean entering into double certification schemes governed by two distinct and separate institutions. Although Fair Trade schemes do have certain sustainability standards, they are not as extensive as those for organic farming. Hence, smallholders who are already under Fair Trade schemes need to adjust and modify their production systems to meet organic standards. This raises the need to explore the possible collaboration of different organizations and players involved in organic farming and Fair Trade certification systems. Hence, this book explores the need and the extent to which such integration could be feasible.

1.4 Organization of the Book

Chapter 2 presents a glance at the organic and Fair Trade markets worldwide; and Chapter 3 discusses the supply trends globally as well as in the developing continents of Africa, Asia and Latin America. The history of producer organizations in Fair Trade systems is presented in detail in Chapter 4. Chapter 5 elaborates a methodological toolbox on fundamental approaches to analysing organic and Fair Trade systems. These include welfare analysis, market studies and social analysis.

Chapters 6–9 discuss consumer case studies from the developed nations of Europe and the USA. Chapter 6 discusses the setting of standards and certification in organic and Fair Trade systems as well as their communication to both producers and consumers. Chapters 7, 8 and 9 investigate the demand side for such certified produce in the developed nations based on country studies in Germany and the USA. These chapters explore consumer preferences for organic and Fair Trade produce as well as willingness to pay, especially for joint organic and Fair Trade commodities in developed nations.

Producer case studies, especially smallholder producers from Africa, Asia and Latin America, are discussed in Chapters 10–15. They are based on empirical evidence and the lessons that can be learnt for successful implementation of these innovations at the producer level. The chapters explore the extent to which these systems are successful in poverty alleviation and providing sustainable livelihoods for smallholder producers in developing countries.

The book concludes with Chapter 16 on the way forward, based on the empirical findings of the different case studies discussed. It raises pertinent questions that remain to be addressed in Fair Trade and organic literature. It assesses the potential merits and shortcomings of joint organic and Fair Trade certification at the consumer and producer level. It further examines the potential integration of different institutions and players involved in organic and Fair Trade systems.

References

Bacon, C.M. (2005) Confronting the coffee crisis: can fair trade, organic, and specialty coffees reduce small-scale farmer vulnerability in northern Nicaragua? *World Development* 33(3), 497–511.

Bacon, C.M. (2008) *Confronting the Coffee Crisis: Fair Trade, Sustainable Livelihoods and Ecosystems in Mexico and Central America*. MIT Press, Cambridge, Massachusetts.

Browne, A.W., Harris, P.J., Hofny-Collins, A.H., Pasiecznik, N. and Wallace, R.R. (2000) Organic production and ethical trade: definition, practice and links. *Food Policy* 25(1), 69–89.

Brundtland Commission (1987) *Our Common Future*. Report of the UN World Commission on Environment and Development. Oxford University Press, New York.

Calo, M. and Wise, T.A. (2005) *Revaluing Peasant Coffee Production: Organic and Fair Trade Markets in Mexico*. Global Development and Environment Institute, Tufts University, Medford, Massachusetts.

Dragusanu, R., Giovannucci, D. and Nunn, N. (2014) The economics of fair trade. *The Journal of Economic Perspectives* 28(3), 217–236.

Geier, B. (2007) IFOAM and the history of the international organic movement. In: Lockeretz, W. (ed.) *Organic Farming: an International History*. CAB International, Wallingford, UK, pp. 175–186.

IFOAM (2006) *The IFOAM Basic Standards for Organic Production and Processing*. Version 2005. Available at: http://agritech.tnau.ac.in/org_farm/pdf/IFOAM_basic_standards.pdf (accessed 5 July 2017).

Krier, J.-M. (2008) *Fair Trade 2007: New Facts and Figures from an Ongoing Success Story*. Dutch Association of Worldshops (DAWS), Culemborg, The Netherlands.

Kumbhakar, S.C., Tsionas, E.G. and Sipiläinen, T. (2009) Joint estimation of technology choice and technical efficiency: an application to organic and conventional dairy farming. *Journal of Productivity Analysis* 31(3), 151–161.

Lansink, A.O., Pietola, K. and Bäckman, S. (2002) Efficiency and productivity of conventional and organic farms in Finland 1994–1997. *European Review of Agricultural Economics* 29(1), 51–65.

Luttikholt, L.W. (2007) Principles of organic agriculture as formulated by the International Federation of Organic Agriculture Movements. *NJAS-Wageningen Journal of Life Sciences* 54(4), 347–360.

Parvathi, P. and Waibel, H. (2013) Fair trade and organic agriculture in developing countries: a review. *Journal of International Food and Agribusiness Marketing* 25(4), 311–323.

Paull, J. (2010) From France to the World: The International Federation of Organic Agriculture Movements (IFOAM). *Journal of Social Research & Policy* 1(2), 93.

Raynolds, L.T. (2000) Re-embedding global agriculture: the international organic and fair trade movements. *Agriculture and Human Values* 17(3), 297–309.

Raynolds, L.T. (2004) The globalization of organic agro-food networks. *World Development* 32(5), 725–743.

Rice, A.R. (2001) Noble goals and challenging terrain: organic and fair trade coffee movements in the global marketplace. *Journal of Agricultural and Environmental Ethics* 14(1), 39–66.

Rigby, D. and Cáceres, D. (2001) Organic farming and the sustainability of agricultural systems. *Agricultural Systems* 68(1), 21–40.

Tzouvelekas, V., Pantzios, C.J. and Fotopoulos, C. (2001) Technical efficiency of alternative farming systems: the case of Greek organic and conventional olive-growing farms. *Food Policy* 26(6), 549–569.

Valkila, J. (2009) Fair Trade organic coffee production in Nicaragua – sustainable development or a poverty trap? *Ecological Economics* 68(12), 3018–3025.

Willer, H. and Kilcher, L. (eds) (2011) *The World of Organic Agriculture. Statistics and Emerging Trends 2011*. Available at: http://orgprints.org/19310/1/world-of-organic-agriculture-2011.pdf (accessed 1 September 2012).

Willer, H. and Lernoud, J. (2016) *The World of Organic Agriculture. Statistics and Emerging Trends 2016*? FiBL, Frick, Switzerland, and IFOAM – Organics International, Bonn, Germany.

2 Organic and Fairtrade Markets at a Glance

Julia Lernoud and Helga Willer*
Department of Extension, Training and Communication, Research Institute of Organic Agriculture (FiBL), Frick, Switzerland

2.1 Introduction

In this chapter, an overview of the global organic and Fairtrade market is presented[(1)]. The data shown here were collected by the Research Institute of Organic Agriculture (*Forschungsinstitut für biologischen Landbau* (FiBL)) in the framework of its surveys on organic agriculture and Voluntary Sustainability Standards (VSS). The organic data are published annually in the statistical yearbook *The World of Organic Agriculture* (Willer and Lernoud, 2017)[(2)] by FiBL together with IFOAM – Organics International. The results of the VSS survey are published in the report *The State of Sustainable Markets* (Lernoud *et al.*, 2017)[(3)] produced in cooperation with the International Trade Centre (ITC) and the International Institute for Sustainable Development (IISD). These efforts are supported by the Swiss State Secretariat for Economic Affairs (SECO). The Fairtrade International data are taken from the *Fairtrade International Annual Report* (Fairtrade International, 2016).

Since 2000, when the global market for organic food was at approximately €16.5 billion (Sahota, 2017), it has more than quadrupled, reaching almost €76 billion in 2015, reflecting a major increase in the demand for organic products in just 15 years. According to Fairtrade International, global Fairtrade sales reached €7.3 billion in 2015.

About 90% of the sales of organic and Fairtrade products are in Europe and North America. For organic, North America is the largest market with over 50% of the global organic market; while for Fairtrade products, Europe represents almost 80% of Fairtrade retail sales (Fairtrade International, 2016).

Data on total organic retail sales value are available for more than 50 countries, most of these being developed countries, which means that such data are missing for many countries with organic farming activities (179 in 2015). For Fairtrade, retail sales data are available for 32 countries, most of them in Europe.

When analysing retail sales data of organic and Fairtrade products, it is important to take into account that, currently, consumers can find almost all food products in organic quality, while the scope of Fairtrade products is smaller. When considering processed foods, another difference emerges between organic and Fairtrade products. For a processed product to be labelled as organic, at least 95% of the ingredients of the final product need to be from organic certified raw materials (IFOAM – Organics International, 2014). In comparison, for a Fairtrade composite food product to carry the Fairtrade label, the

*Corresponding author; e-mail: helga.willer@fibl.org

product needs to have at least 50% of its ingredients Fairtrade certified (Fairtrade International, 2003).

2.2 Highlights and key data

2.2.1 Market size

- The global market for organic food (expressed in value of retail sales) was €75.7 billion in 2015.
- In 2015, Fairtrade retail sales were at €7.3 billion.
- The country with by far the largest market for organic food is the USA (€35.8 billion), followed by Germany (€8.6 billion), France (€5.5 billion) and China (€4.7 billion), as depicted in Fig. 2.1.
- The countries with the largest market for Fairtrade certified products were the UK (€2.2 billion), Germany (€978 million) and the USA (€917 million), as presented in Fig. 2.2.
- Comparing organic markets worldwide by single market, the USA has the lead: 47% of global organic retail sales are in this country (€35.8 billion), followed by the European Union (EU) and China.
- Looking at the Fairtrade retail sales distribution by single market, the EU takes the lead with over 70% of the global market, followed by the USA (almost 13%) and Switzerland (almost 7%).
- Comparing organic retail sales by region, North America continues to be the largest organic market (€38.5 billion), followed by Europe (€29.8 billion) and Asia (€6.2 billion).
- On a regional level, Europe reported the largest market for Fairtrade certified products with almost 80% of the global market, followed by North America with over 16%.
- It is not possible to report a single global figure for Fairtrade and organic, due to the fact that many products are double certified. Fairtrade International reports an up to 60% overlap in some commodities, for example coffee, and for bananas there is a 55% overlap.

2.2.2 Market growth

- The organic market has more than quadrupled in only 15 years, and it has grown each year since 1999, even though a slowdown was noted for some countries during the financial crisis in 2008.
- The market for Fairtrade certified products has grown sixfold since 2004, according to the data collected from Fairtrade International annual reports.

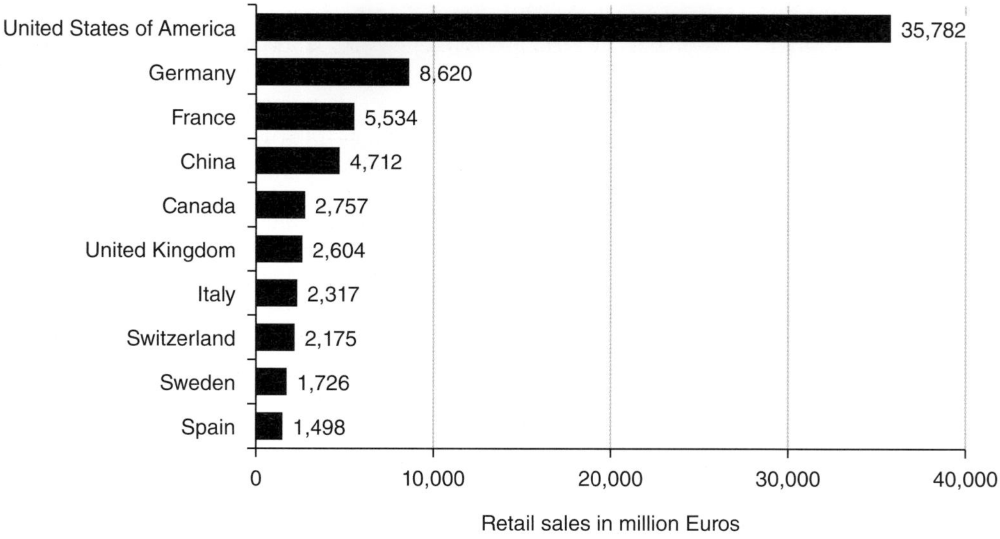

Fig. 2.1. Ten countries with largest markets for organic food, 2015 (FiBL survey 2016: Willer and Lernoud, 2016).

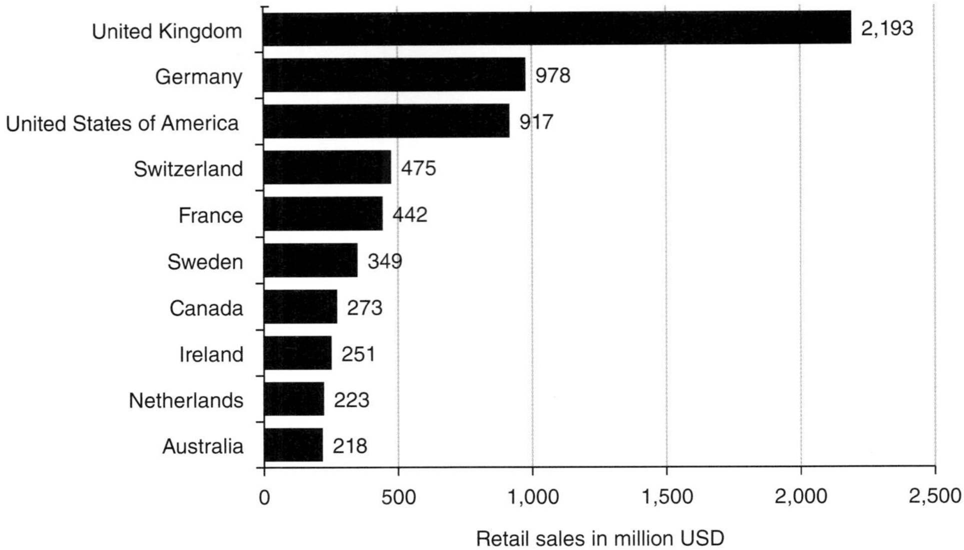

Fig. 2.2. Top ten countries with largest markets for Fairtrade food, 2015 (Fairtrade International, 2015).

- In 2015, organic market growth was noted in all countries, and in some cases, growth was double-digit. For example, in Spain, the market grew by 25%, representing the biggest growth. In Sweden, the market grew by 20%, after having experienced growth of over 40% between 2013 and 2014.
- Retail sales of Fairtrade certified products grew in almost all countries for which data were available. Some countries experienced growth rates of over 20%, such as Estonia (47%), the USA (33%) and Norway (25%).
- Due to fluctuating exchange rates, it is very difficult to calculate a global figure for market growth, but it may be assumed that the global organic market grew by more than 10% in 2015.
- According to Fairtrade International, in 2015 Fairtrade retail sales experienced a growth of 16% compared with 2014.

2.2.3 Per capita consumption

- Globally, €10.3 per person was spent on organic food, and €1 per person was spent on Fairtrade certified products.
- The highest per capita consumption of organic products was registered in European countries. In 2015, Switzerland had the highest per capita consumption worldwide (€262), followed by Denmark (€191) and Sweden (€177).
- Switzerland also had the highest per capita expenditure on Fairtrade certified products in 2015 (€57.7) followed by Ireland (€54.2), Sweden (€36.1) and the UK (€33.9).

2.2.4 Market shares

The global retail sales are presented in Table 2.1. Unfortunately, total retail sales values are not available for all countries.

- Looking at the shares the organic market has of the total market, the leader is Denmark (8.4%), followed by Switzerland (7.7%), Luxembourg (7.5%) and Sweden (7.3%).
- For Fairtrade products, Switzerland and Sweden were the countries with the highest share of the market, accounting for 1.7% and 1.5% of their total market, respectively, followed by Finland (1.3%) and Ireland (1.2%) in 2015.
- 2016 data show that some countries (Denmark, Sweden) are reaching the 10% mark for organic products.

Table 2.1. Global market data: organic and Fairtrade retail sales, share of all retail sales, per capita consumption and market share, 2015. (Fairtrade data: Fairtrade International, *Annual Report 2015–2016*. Organic data: FiBL-AMI survey 2017.) (Note that for market and trade data, comparing country statistics remains very problematic, due to differing methods of data collection. Comments on this table should be sent to helga.willer@fibl.org. Revisions will be posted at www.organic-world.net/statistics/statistics-data-revisions.html and included in the database.)

		Retail sales (million €)		Retail sales (€/person)		Retail sales: growth value 1 year (%)		Retail sales: share value (%)	
Region	Country	Fairtrade	Organic[a]	Fairtrade	Organic[a]	Fairtrade	Organic[a]	Fairtrade	Organic[a]
Africa	Ethiopia		13		0.1				
	Kenya	0.1	4	0.003	0.1	16.0			
	South Africa	19		0.3		–11.0			
Africa total		***19***	***17***						
Asia	Azerbaijan		3		0.3				
	China		4712		3				
	Hong Kong	3		0.4		16.0			
	India	1	130	0.001	0.1	16.0			
	Japan	74	1000	0.6	7.8	7.0		0.1	1.0
	Republic of Korea	5	281	0.1	5.6	16.0			
	Thailand		12		0.2				
	United Arab Emirates		113		12.3				
	Viet Nam		5		0.1				
Asia total		***83***	***6255***	***0.02***	***1.5***				
Europe	Austria	185	1065	21.6	127.0	24.0		1.1	6.5
	Belgium	115	514	10.2	45.7	10.0	18.0	0.6	2.7
	Bosnia and Herzegovina		0.3		0.03				
	Bulgaria		7		0.9				
	Croatia		99		23.4				2.2
	Cyprus		2		1.9				
	Czech Republic	9	74	0.8	7.0	16.0		0.0	0.7
	Denmark	102	1079	18.1	190.7	9.0	12.0	0.80	8.4
	Estonia	3		2.5		47.0			
	Finland	174	240	31.7	43.9	7.0	6.7	1.3	1.8
	France	442	5534	6.7	83.3	13.0	14.6	0.2	2.9
	Germany	978	8620	12.0	105.9	18.0	11.1	0.5	4.8
	Greece		60		5.3				

Continued

Table 2.1. Continued.

Region	Country	Retail sales (million €)		Retail sales (€/person)		Retail sales: growth value 1 year (%)		Retail sales: share value (%)	
		Fairtrade	Organic[a]	Fairtrade	Organic[a]	Fairtrade	Organic[a]	Fairtrade	Organic[a]
	Hungary		30		3.0				
	Ireland	251	142	54.2	30.7	10.0	13.0	1.2	0.7
	Italy	99	2317	1.6	38.1	10.0	15.0	0.1	2.5
	Latvia	1	4	0.4	2.0	–27.0		0.04	0.2
	Liechtenstein		6		142.4				
	Lithuania	1	6	0.3	2.0	25.0		0.02	0.2
	Luxembourg	11	94	19.2	170.0	1.0	4.0	0.9	7.5
	Montenegro		0.1		0.2				
	Netherlands	223	1072	13.2	63.4	4.0	11.5	0.9	4.3
	Norway	81	352	15.6	68.1	25.0	15.0	0.3	1.5
	Poland		167		4.4				
	Portugal	28	21	2.7	2.0	11.0		0.3	0.2
	Romania		80		3.7				0.7
	Russian Federation		120		0.8				
	Slovakia	1	4	0.2	0.7	16.0		0.1	0.2
	Slovenia		49		26.6				1.8
	Spain	28	1498	0.6	32.3	11.0	24.8	0.03	1.5
	Sweden	349	1726	36.1	177.1	19.0	20.3	1.5	7.3
	Switzerland	475	2175	57.7	262.2	9.0	5.2	1.7	7.7
	Turkey		4		0.1				
	Ukraine		18		2.6				
	United Kingdom	2193	2604	33.9	40.2	–5.0	4.9	1.2	1.4
Europe total		***5749***	***29,781***	***7.0***	***36.4***				
Latin America	Belize		0.1		0.2				
	Brazil	0.1		0.001		16.0			
	Chile		2		0.1				
	Costa Rica		1		0.3				
	Mexico		14		0.1				
	Peru		14		0.5				
Latin America total		***0.1***	***31***						

North America	Canada	273	2757	7.6	76.7	16.0		0.3	2.8
	United States of America	917	35,782	2.9	111.2	33.0	11.0	0.1	5.0
North America total		***1189***	***38,539***	***3.3***	***107.7***				
Oceania	Australia	218	962	9.6	41.6	3.0			
	New Zealand		124		27.4				
Oceania total		***218***	***1085***	**5.5**	***27.6***				
Other world		43							
World		**7300**	**75,709**	**1.0**	**10.3**				

[a]Organic retail sales data for the following countries are not from 2015: Australia, 2013; Austria, 2011; Azerbaijan, 2011; Bulgaria, 2010; Chile, 2009; Costa Rica, 2008; Croatia, 2014; Cyprus, 2006; Czech Republic, 2014; Greece, 2010; India, 2012; Japan, 2009; Latvia, 2011; Lithuania, 2011; Mexico, 2013; Montenegro, 2010; Peru, 2010; Portugal, 2011; Romania, 2011; Russian Federation, 2012; Slovakia, 2010; Slovenia, 2013; Thailand, 2014; Turkey, 2009.

Notes

(1) All of the statements and results contained in this chapter have been compiled by the authors and are to the best of their knowledge correct and have been checked by the Research Institute of Organic Agriculture (FiBL). However, the possibility of mistakes cannot be ruled out entirely. The authors are responsible for the content of this chapter. Their opinions do not necessarily express the views of Fairtrade International.
(2) For more information on the global organic survey, please visit http://www.organic-world.net/yearbook/yearbook-2017.html
(3) For more information on the global survey on Voluntary Sustainability Standards, please visit http://www.vss.fibl.org/de/vss.html

References

Fairtrade International (2003) *Guidelines for the Fairtrade-Labelling of Composite Food Products*. Fairtrade International, Bonn, Germany. Available at: https://www.fairtrade.org.uk/~/media/fairtradeuk/what%20is%20Fairtrade/documents/2_comp.pdf (accessed 5 July 2017).

Fairtrade International (2015) *Global Change, Local Leadership*. Annual Report 2014–2015. Fairtrade International, Bonn, Germany. Available at: https://annualreport14-15.fairtrade.net/en/ (accessed 30 June 2017).

Fairtrade International (2016) *Driving Sales, Deepening Impact. Annual Report 2015–2016*. Fairtrade International, Bonn, Germany. Available at: https://annualreport15-16.fairtrade.net/en/ (accessed 4 July 2017).

IFOAM – Organics International (2014) *The IFOAM Norms for Organic Production and Processing – Version 2014*. IFOAM – Organics International, Bonn, Germany. Available at: http://www.ifoam.bio/sites/default/files/ifoam_norms_july_2014_t.pdf (accessed 30 June 2017).

Lernoud, J., Potts, J., Sampson, G., Voora, V., Willer, H. and Wozniak, J. (2017) *The State of Sustainable Markets – Statistics and Emerging Trends 2017*. ITC, Geneva, Switzerland.

Sahota, A. (2017) The global market for organic food & drink. In: Willer, H. and Lernoud, J. (eds) *The World of Organic Agriculture. Statistics and Emerging Trends 2017*. Research Institute of Organic Agriculture (FiBL), Frick, Switzerland; IFOAM – Organics International, Bonn, Germany.

Willer, H. and Lernoud, J. (eds) (2016) *The World of Organic Agriculture. Statistics and Emerging Trends 2016*. Research Institute of Organic Agriculture (FiBL), Frick, Switzerland; IFOAM – Organics International, Bonn, Germany.

Willer, H. and Lernoud, J. (eds) (2017) *The World of Organic Agriculture. Statistics and Emerging Trends 2017*. Research Institute of Organic Agriculture (FiBL), Frick, Switzerland; IFOAM – Organics International, Bonn, Germany.

3 Organic and Fairtrade Production Worldwide

Julia Lernoud and Helga Willer*
Department of Extension, Training and Communication, Research Institute of Organic Agriculture (FiBL), Frick, Switzerland

3.1 Introduction

In this chapter, data on certified organic agriculture and Fairtrade International certified agriculture are presented. For the organic data, the results of the 2016 global organic survey, carried out by the Research Institute of Organic Agriculture (FiBL), are used (Willer and Lernoud, 2016). For Fairtrade, the data as published by Fairtrade International are shown (Fairtrade International, 2015).[(1)] Unless otherwise stated, the source of the data is the above-mentioned publications from FiBL and Fairtrade International. For this chapter, we focus on Africa, Asia and Latin America. However, an overview of the global picture of the organic and the Fairtrade sector is also provided.

For the retail sales, 2015 data were available and were used in this chapter. However, for the other indicators covered in this chapter (area, production, operators, etc.) the latest data available for both standards are from 2014.

3.2 Global overview

3.2.1 Area

In 2014, 43.7 million hectares were under organic agricultural management worldwide and more than 2.4 million hectares were Fairtrade certified (Table 3.1). The region with the most organic agricultural land was Oceania, with 17.3 million hectares, followed by Europe with 11.6 million hectares, Latin America (6.8 million hectares), Asia (3.6 million hectares), North America (3.1 million hectares) and Africa (1.3 million hectares). In contrast, for Fairtrade, the most important region was Africa with 45% of the global Fairtrade certified area, followed by Latin America with 42% and Asia with 13%.

Australia was the country with the most organic agricultural land; Argentina was second, followed by the USA in third place (Fig. 3.1). Peru was the country with the largest Fairtrade certified area (more than 177,000 ha), followed by Côte d'Ivoire (almost 174,000 ha), the United Republic of Tanzania (almost 155,500 ha) and Ghana (more than 155,000 ha) (Fig. 3.2). Some of them were also the countries with the highest proportion of Fairtrade certified area of their total agricultural land.

In addition to the organic agricultural land, there are organic wild collection areas, which constituted more than 37.6 million hectares in 2014. For some regions, such as Africa and Latin America, organic wild collection plays a major role in the organic sector.

The share of the world's agricultural land that is organic was almost 1% in 2014. By region,

*Corresponding author; e-mail: helga.willer@fibl.org

the organic share of total agricultural land was highest in Oceania (4.1%) followed by Europe with 2.4% and Latin America with 1.1%. Many individual countries, however, featured much higher organic shares and in 11 countries more than 10% of the agricultural land was organic. The country with the highest organic share of agricultural land was the Falkland Islands (Malvinas); 36.3% of its agricultural area was organic in 2014 (Fig. 3.3).[2] Compared with 1999, when 11 million hectares were organic, organic agricultural land has almost quadrupled (Willer and Yussefi, 2000) (Fig. 3.4).

Table 3.1. Key indicators for organic and Fairtrade certified agriculture worldwide, 2014 (Fairtrade International 2015 and 2016 and FiBL survey 2016 and 2017).

Indicator	Organic[a]	Fairtrade
Certified agricultural area	43.7 million hectares	2.4 million hectares
Share of total agricultural area	0.99%	0.06%
Producers	2.3 million	1.5 million
Retail sales (2015 data)	€76 billion	€7.3 billion

[a]For organic, the 2015 data were available at the time of writing. In that year, 50.4 million hectares were under organic agricultural management, constituting 1% of the world's farmland, and 2.4 million producers were counted (Willer and Lernoud, 2017).

Almost 0.06% of the world's agricultural land was Fairtrade certified in 2014. There were countries that reach a far higher Fairtrade certified share, such as the Dominican Republic (2.7%), Ghana (1%), Côte d'Ivoire (0.8%) and Peru (0.7%) (Fig. 3.5). Since 2011 (earlier data are not available), when the Fairtrade certified area was 1.9 million hectares, the global Fairtrade area increased by almost 30%, which is more than 0.5 million hectares (Fig. 3.6).

3.2.2 Producers

In 2014, almost 2.3 million organic producers were reported. According to the data obtained, more than three-quarters of the producers were in Asia, Africa and Latin America. The country with the most organic producers was India (650,000, 2013 data), followed by Uganda (190,552) and Mexico (169,703, 2013 data).

There were 1.5 million certified Fairtrade producers in 2014 and there were 1226 producer organizations, in 74 countries. In terms of farmers, the countries with the largest number of farmers were Kenya (almost 400,000), Tanzania (almost 150,000) and Ethiopia (over 140,000).

There were over 800,000 Fairtrade coffee producers, 240,000 tea producers and almost

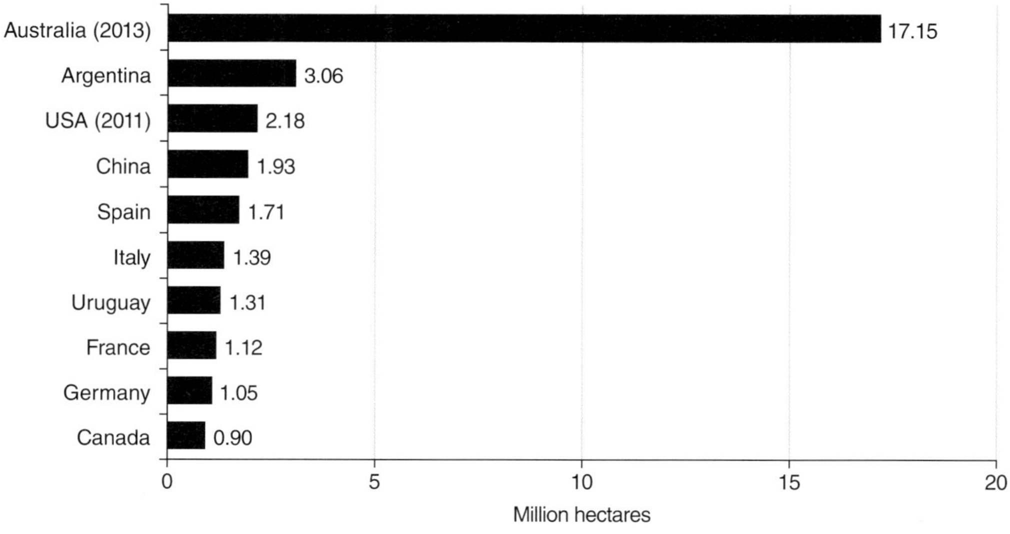

Fig. 3.1. Top ten countries with largest area of organic agricultural land, 2014 (FiBL survey 2016: Willer and Lernoud, 2016).

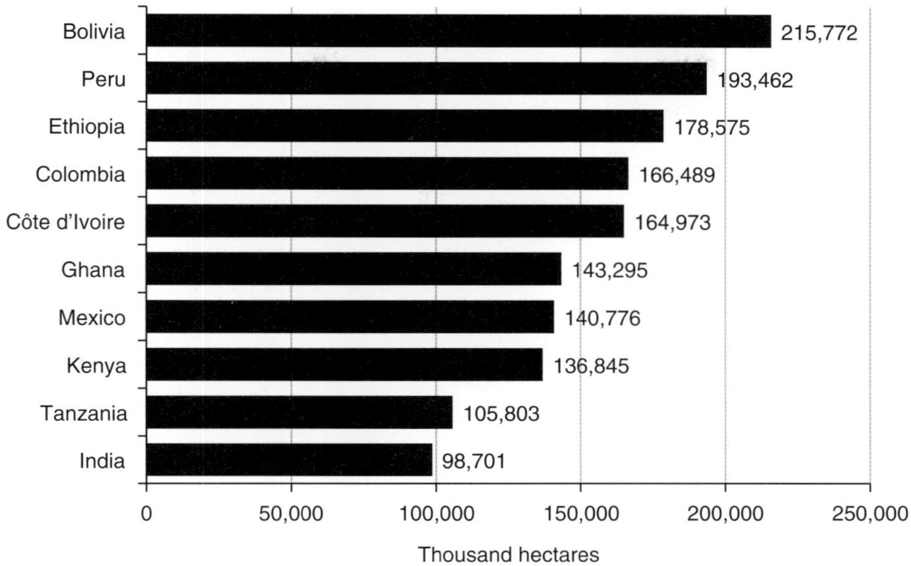

Fig. 3.2. Top ten countries with largest Fairtrade area, 2014 (Fairtrade International, 2015).

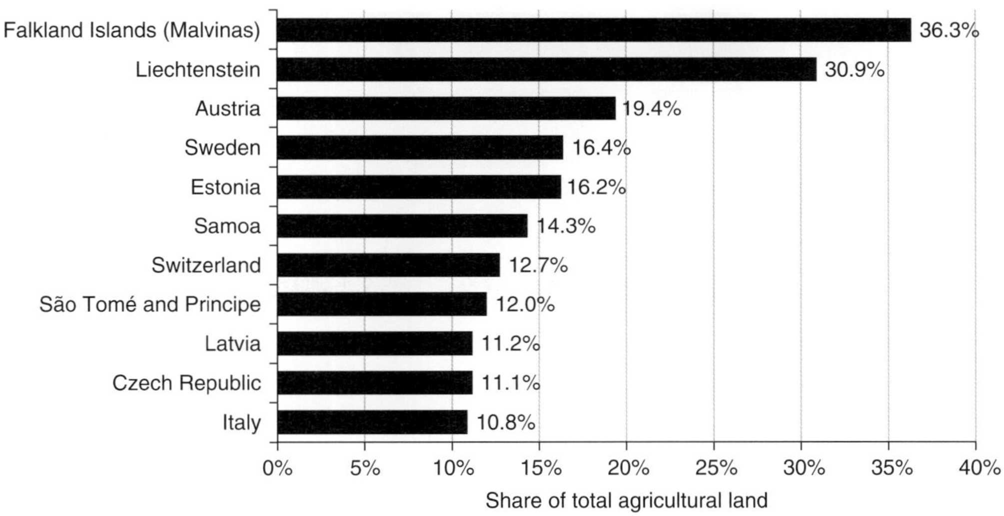

Fig. 3.3. Countries with an organic share of at least 10 per cent, 2014 (FiBL survey 2016: Willer and Lernoud, 2016).

180,000 cocoa producers. Together, they accounted for more than 80% of all of the producers in the Fairtrade system.

According to the Fairtrade International 2015 monitoring report, 52% of all Fairtrade certified producers reported an organic certification and 77% reported holding at least one other certification in addition to Fairtrade (Fairtrade International, 2015). Information on double certification is not available for the area and the production volume, which makes it difficult to report a global area figure for organic and Fairtrade.

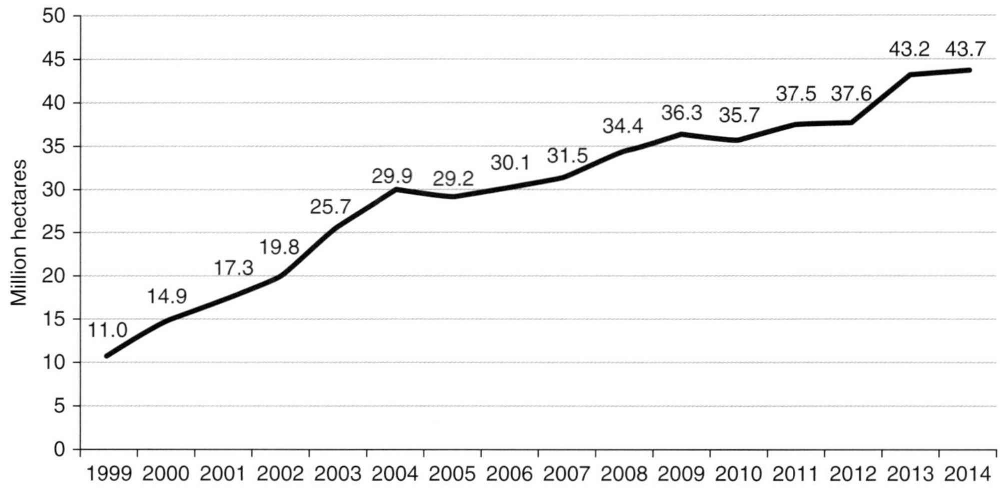

Fig. 3.4. Growth of organic agricultural land, 1999–2014 (FiBL-IFOAM-SOEL Surveys 1999–2016: Willer and Lernoud, 2016).

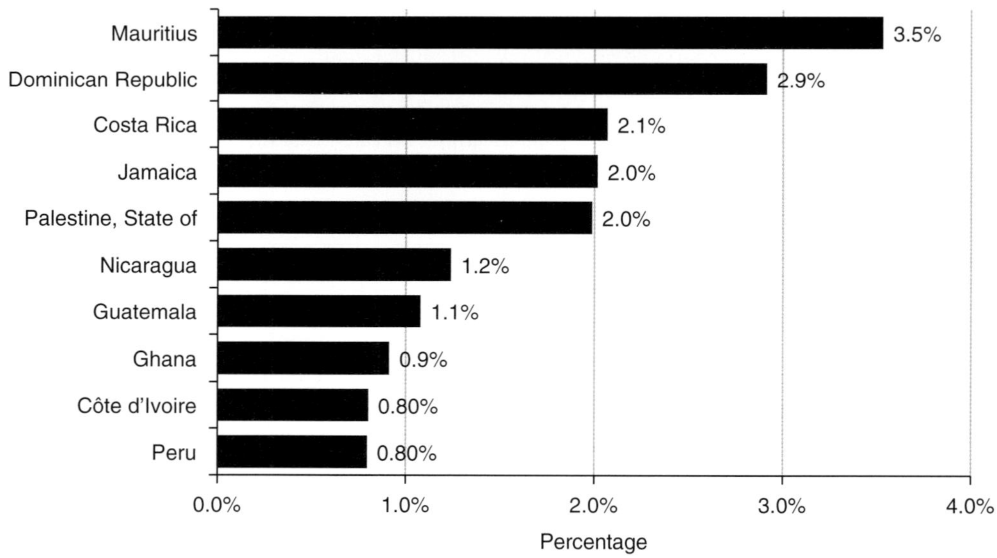

Fig. 3.5. Top ten countries with highest shares of Fairtrade area of total agricultural area, 2014 (Fairtrade International, 2015).

3.2.3 Retail sales

Data on total organic retail sales value were available for more than 50 countries for 2015, which means that for many countries with organic farming activities such data are missing.

In 2015, the countries with the largest market for organic food were the USA (€35.8 billion), followed by Germany (€8.6 billion), France (€5.5 billion) and China (€4.7 billion). The largest single market was the USA, followed by the European Union (€27.1 billion) and China. By region, North America had the lead (€38.5 billion), followed by Europe (€29.8 billion) and Asia (€6.2 billion) (Willer and Lernoud, 2017).

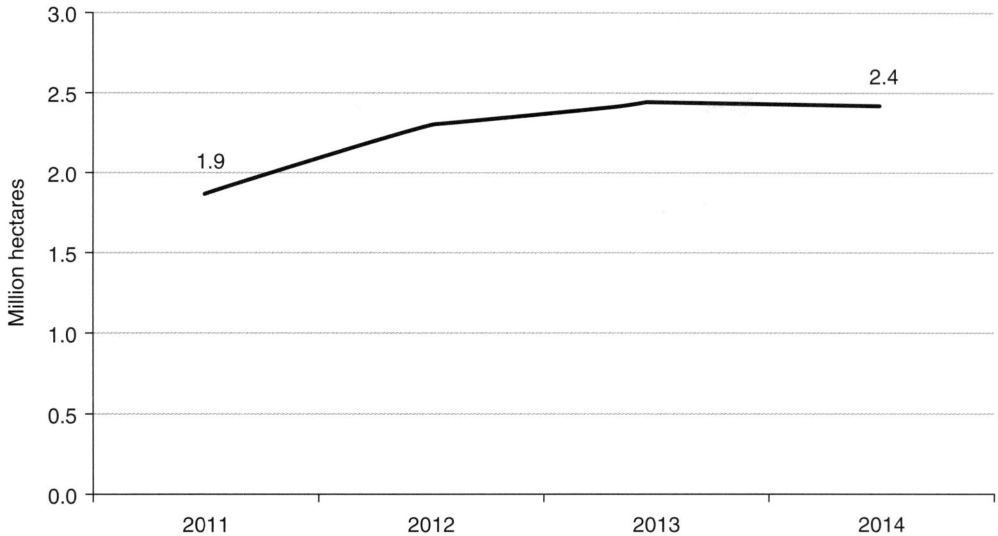

Fig. 3.6. Development of Fairtrade area, 2011–2014 (Fairtrade International, 2015).

In 2015, Fairtrade International retail sales were at €7.3 billion and the largest markets were the UK (€2.2 billion), Germany (€978 million) and the USA (almost €917 million) (Fairtrade International, 2016).

In 2015, organic market growth was noted in all countries and in some cases the market grew at double-digit rates; in Spain, for example, the market grew by more than 25%, representing the biggest growth (Willer and Lernoud, 2017). According to Organic Monitor, the organic market has grown by over 400% since 1999 (Sahota, 2017).

According to Fairtrade International (2016), Estonia was the country that in 2015 reported the highest retail sales growth of Fairtrade certified products, an increase of 47% compared with 2014. Close to Estonia came the USA with a growth of 33% and Norway with 25%. Since 2004, when the global Fairtrade market was over €800 million, it had grown sixfold.

The highest organic per capita consumption was in European countries: in 2015, Switzerland had the highest organic per capita consumption (€262) worldwide, followed by Denmark (€191) and Sweden (€177). Looking at the shares the organic market had of the total market, the leader was Denmark (8.4%), followed by Switzerland (7.7%), Luxembourg (7.5%) and Sweden (7.3%) (Willer and Lernoud, 2017).

Switzerland reported the highest per capita expenditure in Fairtrade certified products (€57.7) in 2015. Close to Switzerland came Ireland with €54.2 per capita, followed by Sweden (€36.1) and the UK (€33.9) (Fairtrade International, 2016).

It is worth noting that in 2015 both the organic and the Fairtrade certified retail sales had high growth in different European countries, showing a growing interest of consumers in sustainable and ethical produce. Furthermore, Switzerland was the country with the largest consumption for both organic and Fairtrade certified products.

3.2.4 Exports

International trade data are becoming available for more and more countries. These can be expressed as total export/import volumes in metric tons (t) or as values. Some countries also provide a breakdown by product. However, it still remains difficult to draw a global picture on the flow of organic goods. From the 178 countries with organic farming, export data were only available for 40, i.e. only 20% (Willer and Lernoud, 2017).

No data on export of Fairtrade certified commodities exist. However, producers' reports of their sales in 2013–2014 indicated strong growth of the volumes of bananas, cocoa and seed cotton sold as Fairtrade: 15%, 17% and 21%, respectively, according to the Fairtrade International 2015 Report. Other major products, including coffee, sugar and flowers, registered more moderate growth in volumes sold on Fairtrade terms, at 6%, 4% and 5%, respectively.

The three largest Fairtrade products by farm-gate sales value – coffee, bananas and cocoa – generated 80% of the reported Fairtrade sales income for producers. Furthermore, 40% of all producer organizations sold more than half of their production volume on Fairtrade terms in 2013–2014 (Fairtrade International, 2015).

3.3 Africa

3.3.1 Overview

There is a growing recognition among policy makers that organic agriculture has a significant role to play in addressing food insecurity, land degradation, poverty and climate change in Africa. IFOAM – Organics International is currently working with the African organic sector, the African Union and other agencies in the framework of its 'Organic Alternative for Africa Initiative' to facilitate the integration of organic agriculture into the core of African policies and the agricultural development agenda. Furthermore, in 2011 the African Union published its decision to support the development of the organic sector in the region, through the creation of a coalition with international partners to provide guidance on sustainable organic farming systems and improvement of seed quality (African Union, 2011).

Significant breakthroughs were achieved in 2013 with the institutionalization of the African Organic Network (AfrONet). Its activities include strengthening and supporting regional networks, the Ecological Organic Agriculture Initiative for Africa (EOA, 2015), and organic conferences in Eastern, Western, Central and Southern Africa (Gama, 2015).

Since 2013, Fairtrade Africa, the regional body of Fairtrade International in Africa, has been implementing a programme aimed at improving producers' access to finance in five countries (Kenya, Tanzania, Uganda, Ghana and Côte d'Ivoire). Producer organizations are being trained in financial management and mentored in finance by experts hired through the project (Fairtrade International, 2015).

In March 2014, Ferrero announced a commitment to purchase 20,000 t of Fairtrade certified cocoa from Côte d'Ivoire over 3 years. This is the largest Fairtrade Sourcing Program (FSP) commitment to date and a total of 18 companies have begun to source Fairtrade cocoa through this programme. Thanks to the commitment from Ferrero, producers in Côte d'Ivoire were able to sell more of their cocoa as Fairtrade in 2014 as a result of the FSP for cocoa (Fairtrade International, 2015).

Furthermore, Fairtrade Africa commissioned AidEnvironment, a consultancy organization focusing on sustainable production and trade, to undertake research with Fairtrade cotton farmers in West Africa. Researchers collected data from Fairtrade certified and non-Fairtrade certified farmers from different producer organizations in Senegal, Mali and Burkina Faso. The research aimed to provide a strong basis for the evaluation of the impacts of Fairtrade cotton certification in West Africa. The study found that if the costs of non-paid labour were not taken into account, Fairtrade farming was more cost-efficient than non-Fairtrade (and Fairtrade organic farming even more so) (Fairtrade International, 2015).

In Africa there were almost 1.3 million hectares of organic agricultural land in 2014, which is 0.1% of the region's total agricultural area and 3% of the global organic agricultural area (Table 3.2). In 2014, 39 African countries reported data on organic farming. The area of organic agricultural land has increased by more than 1 million hectares from 52,000 ha in 2000. Uganda was the country with the largest organic area, with more than 240,000 ha and with the largest number of organic producers. The country with the highest proportion of organic agricultural land was the island state of São Tomé and Principe, with 12% of its agricultural area being organic, followed by Egypt with 2.3%, Uganda with 1.7% and Réunion with 1.6%.

In 2014, there were almost 870,000 ha of Fairtrade certified land in Africa, representing 0.07% of the region's total agricultural area.

Table 3.2. Key indicators: organic and Fairtrade certified agriculture in Africa, 2014 (Fairtrade International 2015 and FiBL survey 2015 (Willer and Lernoud, 2016)).

Indicator	Organic	Fairtrade
Certified agricultural area	1.3 million hectares	870,000 ha
Share of total agricultural area	0.1%	0.07%
Producers	590,000	Over 1 million
Key commodities (based on area)	Coffee, olives and cotton	Coffee, cocoa and cotton

Overall growth of the number of producers in both percentage and absolute terms was strongest in Africa, where their number grew by almost 123,000, representing an overall growth rate of 13%. Sixty-four per cent of Fairtrade certified producers are in Africa (over 1 million) (Fairtrade International, 2015).

By the end of 2014, there were 392 Fairtrade certified producer organizations in 32 countries throughout Africa. Of these, 288 were small producer organizations and 104 were hired labour organizations (Fairtrade International, 2015).

There were at least 590,000 organic producers in Africa. The countries with the most organic producers were Uganda (190,000), the United Republic of Tanzania (148,000) and Ethiopia (almost 136,000). It can be assumed that the number of producers was higher, because some countries only reported the number of farm enterprises/companies. Since 2004, when almost 96,000 organic producers were counted, their number increased fivefold.

3.3.2 Land use and key crops

In 2014, almost half of all organic farmland in Africa was used for permanent crops (more than 600,000 ha), almost 20% was used for arable crops (almost 244,000 ha) and almost 6% (71,000 ha) was grassland/grazing area. For 28% of the agricultural land, no land use or crop details were available.

Ethiopia (154,000 ha, mainly coffee), Tunisia (135,000 ha, mainly olives), the United Republic of Tanzania (over 124,000 ha) and the Democratic Republic of the Congo (40,000 ha) had the largest organic permanent crop areas. The key organic permanent crop was coffee, amounting to at least 223,000 ha, which was 12% of Africa's total coffee area. As no crop details were available for some of the largest African coffee producers (e.g. Uganda), it can be assumed that the total figure for organic coffee was higher. The largest organic coffee areas were in Ethiopia and the Democratic Republic of Congo (almost 26,000 ha). The organic coffee area had increased tenfold since 2004. Cocoa was grown on almost 39,000 ha and the area had grown 15-fold since 2004, representing 0.6% of the continent's cocoa area. The largest areas of organic cocoa were found in the Democratic Republic of Congo (almost 14,400 ha), São Tomé and Principe (6401 ha) and Ghana (5153 ha).

Almost 20% of the organic farmland was used for arable crops, most of which were oilseeds (mainly sesame; almost 124,000 ha, i.e. 0.6% of the total oilseed area) and cotton (almost 68,000 ha, i.e. 1.7% of the total cotton area) and aromatic and medicinal plants. Over 65% of the organic cotton area was in the United Republic of Tanzania (more than 44,000 ha), followed by Sudan (15,000 ha). Since 2004, the organic cotton area had grown by over 600%. Aromatic and medicinal plants were grown on over 20,800 ha in 2014; the key producing countries were the United Republic of Tanzania (10,000 ha), Madagascar (almost 5000 ha) and South Africa (almost 3000 ha). The area devoted to aromatic and medicinal plants had grown 24-fold since 2004; however, some of the increase must be attributed to the continually improving availability of data.

Wild collection has an important role in Africa, with almost 11.5 million hectares certified as organic in 2014. Zambia was the country with the largest area, with almost 7 million hectares (mainly beekeeping), followed by Namibia (2.4 million hectares, medicinal plants), Morocco (almost 862,000 ha, mainly argan oil) and Zimbabwe (549,600 ha, mainly devil's claw (*Harpagophytum procumbens*)). Medicinal plants such as devil's claw played the most important role in wild collection, with almost 3 million hectares, followed by oil plants (over 800,000 ha) such as shea nuts (almost 102,000 ha). Beekeeping was the key activity in organic wild collection in Africa, with 6.6 million hectares.

Zambia was the country with the largest wild collection area used for organic beekeeping, with over 6 million hectares, representing 98% of the organic beekeeping area, followed by Cameroon (360,000 ha), Kenya (89,000 ha) and Nigeria (1000 ha).

Coffee was the largest Fairtrade cultivated commodity in Africa in 2014 (over 356,000 ha), representing almost 19% of the total coffee area in the region. Ghana was the country with the largest area (almost 1400 ha) and produced almost 51,000 t. Close to coffee, Fairtrade cocoa was also largely cultivated (331,000 ha) and covered over 5% of the total cocoa area in the region.

Coffee, cocoa and cotton play a significant role in Africa for organic production and Fairtrade certified production. From the data available, these cash crops are widely grown and involve a large number of producers.

3.3.3 Retail sales

Although the African market for organic products is still small, domestic organic markets are growing in Africa. Organic markets and specialized shops are usually located near capital cities and the majority of the consumers are foreigners and upper-middle-class citizens with values similar to those of European organic consumers. Organic shops in South Africa, Kenya, Uganda and Ghana are picking up organic products. In Zambia, organic farmers sell their produce at local farmers' markets or to urban supermarkets. The main organic marketed products are fresh fruit and vegetables, dairy products, meat, wine, herbs and personal care products (Bouagnimbeck, 2011).

With increasing awareness, it can be assumed that the potential of local or domestic African markets for organic products will increase. However, few African countries have articulated a concrete promotion strategy for domestic markets (Bouagnimbeck, 2011).

Fairtrade Africa is supporting women's participation and representation in local coffee production networks through the establishment of a women's coffee association and a learning platform. The coffee grown by these women will be branded and sold in domestic markets (Fairtrade International, 2015). For Africa, Fairtrade retail sales data for 2015 are available for Kenya (€0.1 million) and South Africa (€19 million) (Fairtrade International, 2016) and organic retail sales data are available for Ethiopia (€13 million) and Kenya (€4 million) (Willer and Lernoud, 2017).

3.3.4 Exports

The majority of certified organic products in Africa are destined for export markets. The European Union, as the major recipient of these exports, is Africa's largest market for agricultural products (Bouagnimbeck, 2011).

In most cases, due to the dominance of smallholders in Africa, the typical supply chain comprises a private enterprise organizing many smallholders as out-growers to secure sufficient quantities for export, or farmers work together on one project supplying and packaging for exporting trading companies (Bouagnimbeck, 2011).

Data on organic exports are only available for Ethiopia and Uganda. Ethiopia reported €144 million (mainly coffee) and Uganda €34 million (Willer and Lernoud, 2017). Export data for Fairtrade certified products are not available.

3.4 Asia

3.4.1 Overview

Consumer demand for organic food is growing in Asia, partly due to consumer concerns about food safety, which has been a major driver of organic food sales in China. Community Supported Agriculture (CSA) and Participatory Guarantee Systems (PGS)[(3)] are developing apace as low-cost alternatives to third-party certification (Ong, 2015).

Fairtrade International is working to further develop the sector in Asia by organizing workshops, research and training. Fairtrade International is working with a wide range of partners to support improved wages for workers on tea estates, including participation in industry-wide initiatives such as the Oxfam Tea Wage Report (Oxfam, 2013) and Tea 2030 (Forum for the Future, 2014).

Table 3.3. Key indicators: organic and Fairtrade certified agriculture in Asia, 2014 (Fairtrade International 2015 and FiBL survey 2016 (Willer and Lernoud, 2016)).

Indicator	Organic	Fairtrade
Certified agricultural area	3.6 million hectares	260,000 ha
Share of total agricultural area	1.3%	0.02%
Producers	900,000 organic producers	300,000 Fairtrade certified producers
Key commodities (based on area)	Cereals, oilseeds, and coconuts	Coffee, tea, cotton and sugarcane

The organic agricultural land in Asia amounts to almost 3.6 million hectares, which is 0.3% of the total agricultural area in the region (Table 3.3). Eight per cent of the global organic agricultural land is in Asia. Compared with 2001 (420,000 ha), organic land has increased almost eightfold. Between 2013 and 2014, the organic area in Asia increased by 158,500 ha, or 4.7%, continuing to recover after the loss of half a million hectares in India in 2012. The country with the largest organic agricultural area was China (1.9 million hectares) and the country with the most producers was India (650,000 producers in 2013). The countries with the highest organic shares of the total agricultural land were Timor-Leste (6.8%) and Sri Lanka (2.3%).

In 2014, 900,000 organic producers were reported in Asia. After India, the Philippines had the second most producers (almost 166,000). Many countries do not report the number of producers or only report the number of companies, and so it can be assumed that the number of producers is higher. Since 2004, when there were 100,000 organic producers, the number increased almost eightfold.

Almost 260,000 ha were Fairtrade certified in Asia in 2014. This represented 0.02% of the region's agricultural area.

In 2014, there were 187 Fairtrade certified producer organizations in 18 countries in the region, constituting a growth of 3% compared with 2013. Fairtrade certified producers in Asia (almost 300,000) represented 16% of all Fairtrade certified producers.

3.4.2 Land use and key crops

In 2014, 45% of all organic farmland was used for arable crops (1.6 million hectares), 1% for grassland/grazing areas (almost 28,000 ha) and 15% (541,000 ha) for permanent crops. Detailed information was not available for 37% of the agricultural land, so it can be assumed that each category has a far larger share of the total organic land.

The key organic arable crop group was cereals (mainly wheat and rice), with over 755,000 ha reported in total, representing 0.2% of the total cereal area in Asia. Most organic cereals were grown in China (almost 566,000 ha) and Kazakhstan (130,000 ha). Oilseeds (mainly soybeans) were also an important crop group, grown on at least 443,000 ha, mainly in China and India, and represented 0.8% of the total oilseed area in Asia. The key organic cereals were wheat, rice and grain maize. Organic wheat represented over 30% of the total organic cereal area and 0.3% of the total cereal area in Asia. The largest organic wheat areas were in China (almost 121,500 ha) followed by Kazakhstan (almost 120,000 ha). Organic rice was mainly grown in China (almost 181,500 ha,) constituting 80% of the total organic rice of the region.

Almost 15% of the organic farmland was used for permanent crops; most of this land was used for coconuts (almost 122,000 ha), coffee (113,000 ha) and tea (at least 58,000 ha). The Philippines was the country with the largest area of organic coconuts with 80,500 ha, representing over 66% of the total organic coconut area of the region. Most of the organic coffee grown in Asia was found in Indonesia, where 81,500 ha were reported, followed by Timor-Leste (more than 25,000 ha); the two countries represented almost 95% of the organic coffee area in Asia. Organic coffee represented 4.4% of the total coffee area in Asia. Almost 2% of the total tea grown in Asia was organic; most of it was in China (46,000 ha), followed by Myanmar (4000 ha).

In 2014, 6.3 million hectares of organic wild collection were reported in Asia. No detailed data are available for 87% of the reported area but, from the details available, wild fruits and wild mushrooms were the key commodities; furthermore wild oil plants (44,700 ha) and medicinal plants (18,400 ha) played an important role. More than 0.4 million hectares of wild fruits were reported and more than 91,000 ha of wild mushrooms. India is the country in the region with the largest organic wild collection area, with almost 4 million hectares, followed by China (1.14 million hectares) and Tajikistan (1.06 million hectares).

In Asia, coffee was the most important Fairtrade certified commodity; it was grown on over 75,000 ha, representing almost 3% of the region's coffee area. Fairtrade sugarcane, cotton and tea were also important commodities. They covered almost 66,000 ha (over 55,000 ha and 37,000 ha, respectively). Fairtrade cotton represented almost 0.3% of the regional cotton area. India had the largest Fairtrade cotton area, representing over 90% of the total Fairtrade cotton area (almost 53,000 ha). Furthermore, India was the country with the largest Fairtrade tea area (over 21,000 ha).

Coffee and tea play a significant role in Asia in organic and Fairtrade certified production. From the data available, these cash crops are widely grown and involve a large number of producers. However, in organic farming coconut and rice are the largest commodities grown; and for Fairtrade, sugarcane and cotton are also very important.

3.4.3 Retail sales

Domestic markets based on organic PGS are growing. Three national associations were formed, in Nepal, Lao People's Democratic Republic and Vietnam (Ong, 2013). Globally, Asia has the third largest market for organic products (after North America and Europe). Although the continent has over 3 million hectares of organic agricultural land, many organic food products are imported. China produces many organic crops for domestic consumption; however, other countries mainly export primary crops. Most countries, including Japan, Singapore, Taiwan and South Korea, have import-dependent organic food sectors (Sahota, 2016).

In India, the organic domestic market is growing and now has over 2000 organic shops and markets (Ong, 2013). In Saudi Arabia, retailers specializing in organic and health foods are gaining importance (Hartmann *et al.*, 2013).

In Vietnam, organic has received high press coverage and domestic demand for organic products, particularly tea and vegetables, is growing. Many organic retail shops handling tea and vegetables have opened in Hanoi, the capital city of Vietnam (Ong, 2016).

A study conducted in 2015 by the Organic Development Center (Sukhothai Thammathirat Open University) and the Earth Net Foundation showed that in Thailand, within the domestic market, supermarkets were the most important sales channel of the total domestic sales, followed by 'green' shops and restaurants, cafés and food services (Panyakul, 2016).

In Bangladesh, organic commodities are often sold via specific sections of many conventional department stores. Other marketing channels include contract farming and direct sales from farmers at local markets (Saha and Monowar Hossain, 2014).

In Asia, organic market data are not available for all countries, but it can be assumed that the market is continually growing. Eight countries provided organic retail sales values in 2015 (Willer and Lernoud, 2017). For China, values of €4.7 billion were reported for 2015, making the country the world's fourth biggest market for organic products. Japan has a large organic domestic market valued at €1 billion and South Korea reported a market of €281 million.

In 2015, Fairtrade retail sales data were available for four countries (Fairtrade International, 2016). Japan reported the largest market with €74 million, an increase of 7% compared with 2014, followed by South Korea with €5 million (16% increase), Hong Kong with €3 million (16% increase), and India with €1 million (16% increase). The four countries showed a high growth compared with 2014; all countries reported double-digit growth, with the exception of Japan.

For both organic and Fairtrade certified retail sales, Japan and South Korea represent important markets for the region. Fairtrade certified retail sales data for China are not available.

3.4.4 Exports

As in Africa, most of the organic production in Asia is meant for export. However, China and Japan import much of their demand from within the region. Europe and North America remain key recipients of Asia's organic exports. Key organic export commodities are rice, tea, tropical fruits and spices, among others.

In 2014, Vietnam reported the largest value of organic exports (€551 million). China reported almost €470 million of organic exports, India reported €303 million, Sri Lanka over €170 million and Thailand almost €30 million. To collect export and import data remains difficult and it can be assumed that some of these figures are incomplete; data for many organic producing countries are missing.

3.5 Latin America

3.5.1 Overview

Many Latin American countries remain important exporters of organic products, such as bananas, cocoa and coffee; in countries like Argentina and Uruguay, temperate fruit and meat are key export commodities. At the same time, organic domestic markets are trending positively in the region. Healthy products and the gastronomy sector have also been drivers of the organic sector in many countries with value-added products and visible marketing processes, such as highly nutritious or gourmet fruits and vegetables. PGS are gaining more recognition among consumers (Flores and Soberanes, 2016).

Climate change impacts and the expansion of coffee rust disease pose major challenges for coffee farmers in Latin America. In 2014, Fairtrade International and the Producer Network for Latin America and the Caribbean (CLAC) launched a pilot project to recover coffee bushes affected by coffee rust in El Salvador. The project fosters collaboration and learning between producer organizations from different countries in Central America (Fairtrade International, 2015).

Fair Trade America is working on innovative ways to increase the productivity of smallholder banana farmers. These approaches require limited additional investment and use methodologies that are adapted to the needs of small-scale farmers (Fairtrade International, 2015).

In 2014, 6.8 million hectares were reported as being under organic production, which is 1.1% of the total agricultural land in Latin America and the Caribbean (Table 3.4). Sixteen per cent of the world's organic agricultural land was in Latin America and the Caribbean. Almost 30,000 ha less were reported compared with in 2013. This can be partly attributed to a major decrease, mainly of grassland/grazing areas, in Argentina (over 200,000 ha less). However, in some countries there was a major increase of organic agricultural land in 2014. The organic area has increased by over 70% since 2000 (3.9 million hectares). The country with the largest organic agricultural area was Argentina, with 3 million hectares, and the country with the largest number of producers was Mexico, with more than 169,000 producers.[(4)]

Table 3.4. Key indicators: organic and Fairtrade certified agriculture in Latin America and the Caribbean, 2014 (Fairtrade International 2015 and FiBL survey 2016 (Willer and Lernoud, 2016)).

Indicator	Organic	Fairtrade
Certified agricultural area	6.8 million hectares	818,367 ha
Share of total agricultural area	1.1%	0.1%
Producers	380,000 organic producers	647 Fairtrade certified producers
Key commodities (based on area)	Coffee, cocoa, and tropical and subtropical fruits	Coffee, cocoa, and sugarcane

In Latin America there were at least 380,000 organic producers in 2014. Mexico was the country with the most organic producers (almost 170,000), followed by Peru (more than 65,000), Paraguay (58,000) and the Dominican Republic (more than 26,000). Many countries do not report the number of producers, or they only report the number of companies, and so it can be assumed that the number of producers is higher in Latin America. Since 2004, when there were almost 130,000 organic producers, their number has doubled.

At the end of 2014, there were 647 Fairtrade certified producer organizations in 24 countries in Latin America and the Caribbean, a 2% growth compared with 2013. Of these, 572 were small producer organizations and 75 were plantations. The major growth in new Fairtrade producer organizations was in Peru, where in particular coffee-producing organizations joined Fairtrade. Latin American and Caribbean Fairtrade certified small-scale farmers and workers (over 300,000) represent 20% of the Fairtrade certified producers worldwide (Fairtrade International, 2015). Almost 820,000 ha were Fairtrade certified in 2014, representing 0.13% of the agricultural area of the region.

3.5.2 Land use and key crops

Land use details were available for more than 80% of the region's organic agricultural land. In 2014, only 5% of all organic farmland was utilised for arable crops (almost 328,000 ha), while almost 70% was grassland/grazing areas (4.5 million ha) and 12% (almost 798,000 ha) was used to grow permanent crops. For 16% of the reported area no details were available. Argentina (2.8 million hectares), Uruguay (1.3 million hectares) and the Falkland Islands/Malvinas (0.4 million hectares) had the largest permanent grassland/grazing areas.

The key organic arable crops were cereals, with 123,000 ha, representing almost 40% of the organic arable area of Latin America and Caribbean and 0.2% of the total cereal area in the region. Most of the cereals were grown in Bolivia (87,000 ha, mainly quinoa and amaranth), Argentina (21,354 ha, mainly wheat) and Peru (6000 ha, mainly quinoa). The key organic cereal in the region was quinoa (almost 95,000 ha) constituting more than 70% of all the quinoa grown in the region. Organic sugarcane was grown on more than 61,000 ha in 2014, which was 0.5% of the total sugarcane in the region. The key producing countries were Paraguay (40,000 ha) and Argentina (11,000 ha).

The main organic permanent crops were coffee (over 400,000 ha), cocoa (206,000 ha) and tropical and subtropical fruits (almost 124,000 ha). Organic coffee represented almost 8% of the total coffee area in the region and more than half of the world's organic coffee. The countries with the largest organic coffee areas were Mexico (almost 243,000 ha), Peru (more than 89,000 ha) and Honduras (23,500 ha). Furthermore, 12% of the cocoa area in Latin America was organic, representing over 80% of the world's organic cocoa area, and the countries with the largest organic cocoa areas were in Latin America. The Dominican Republic was the country with by far the largest area, with 120,000 ha (almost 50% of the global organic cocoa area), followed by Peru (more than 25,000 ha) and Mexico (almost 19,400 ha). Organic bananas were the main tropical fruit grown in the region (almost 54,000 ha) representing 2.5% of the regional banana area. The countries with the largest organic banana area were the Dominican Republic (25,000 ha) and Ecuador (17,000 ha); these two countries represented almost 80% of the regional organic banana area.

Wild collection plays an important role in Latin America and the Caribbean. There were more than 3 million hectares of organic wild collection areas in 2014. They were mainly used for the collection of nuts (more than 1 million hectares), palmito (almost 64,000 ha), rose hips (58,000 ha) and berries (almost 18,000 ha). The countries with the largest organic wild collection areas were Brazil (1.2 million hectares, data 2011), Bolivia (0.9 million hectares) and Argentina (0.5 million hectares). Information on wild collection is not available for many countries and it can be assumed that the total wild collection area is higher than that reported here.

Coffee is the largest Fairtrade certified commodity grown in Latin America with over 580,000 ha, representing almost 11% of the region´s coffee area and 64% of the total Fairtrade coffee area. Fairtrade cocoa and sugarcane are also important, accounting from almost 92,000 ha and over 73,000 ha, respectively. Fairtrade cocoa represented 5.4% of the regional cocoa area. The Dominican Republic had the largest area, with almost half of the total Fairtrade cocoa area (almost 46,000 ha). Fairtrade bananas were grown on over 31,000 ha and represented 1.5% of the region's banana area.

As in Africa and in Asia, coffee plays a significant role in Latin America for both organic and Fairtrade certified production. Cocoa and sugarcane are important commodities for both labels. However, in the south of the region, organic pome fruits, such as apples and pears, and organic meat represent a large share of the regional organic production.

3.5.3 Retail sales

Organic domestic markets are trending positively in the region. Today, the public is more environmentally aware and has greater purchasing power than in the past, and there are more producers supplying organic food in alternative commercial schemes, such as fairs, home deliveries and small food stores. For some organic producers in the region, the domestic market is more attractive than the international market, because of requirements and compliances that have to be met (Flores, 2015).

In Latin America, Brazil has the largest market for organic products. As in Europe, most turnover is by large food retailers. Private labels are also slowly gaining popularity in the region. Argentina, Peru, Chile and Colombia are important producers and exporters of organic crops (Sahota, 2016). In Brazil, major supermarket chains offer their own organic and health brands (Flores, 2014).

In Mexico, the Tianguis Network and Organic Markets (*Red de Tianguis y Mercados Organicos*) has developed an outstanding decentralized network of fairs throughout the Mexican territory (Flores and Soberanes, 2016).

In Peru, the organic domestic market has grown steadily in the past 5 years; in Lima, there are 15 different organic street fairs every weekend (Flores and Soberanes, 2016).

In Honduras, a Fairtrade certified coffee-producing organization has invested in organic vegetable production and runs a weekly market, where members can sell their vegetables (Fairtrade International, 2015). Fairtrade retail sales data for 2015 are available only for Brazil (€0.1 million) and no data are available for other Latin American countries (Fairtrade International, 2016).

3.5.4 Exports

Although the domestic market is developing, organic production in Latin America is still export oriented. Apart from tropical fruits and cash crops such as cocoa or coffee from many countries in the region, apples and meat from Argentina, quinoa from Bolivia and sugarcane from Paraguay are key commodities that the region provides to the developed markets in the north. The USA and some European countries are the main recipients of these commodities. Latin America exports mainly raw materials that are then processed into final products in the importing country, where they will be sold to the consumer. For example, cocoa beans are exported to Italy, where they are transformed into chocolate.

In 2014, Peru reported €255 million of organic exports (mainly coffee and cocoa), the Dominican Republic €172 million (mainly cocoa) and Chile reported €154 million (mainly berries and wine). Data on Fairtrade certified exports are not available for this region.

3.6 Conclusions

From the available data in the 2015 edition of *The State of Sustainable Markets*, we conclude that organic farming is the largest sustainable farming method used worldwide. No other standard has a higher number of producers, a higher retail sales value or a larger certified area. Fairtrade certified agriculture is growing at a steady pace and area and retail sales are growing faster than for organic. However, whereas Fairtrade International only certifies selected commodities, almost all agricultural commodities are available from certified organic production.

Reporting a global total of a given commodity or of the whole sector remains difficult as many producers are multiple-certified by both standards and there are not enough reliable data on the share of multiple certifications.

Notes

(1) The Fairtrade data were collected in the framework of the global survey on Voluntary Sustainability Standards (VSS) 2015 and 2017. This survey covered 14 standards from different sectors and was carried out in cooperation with the International Trade Centre (ITC) and the International Institute for Sustainable Development (IISD) (Lernoud *et al.*, 2015, 2017). Both the organic and the VSS surveys were supported by the Swiss State Secretariat for Economic Affairs (SECO).

(2) In 2015, a couple of farmers with large areas of organic grassland did not renew their organic certification. Currently (2015 data), Liechtenstein has the largest organic share of the total farmland (more than 30%).
(3) For more information on Participatory Guarantee Systems (PGS), please visit www.ifoam.bio/en/organic-policy-guarantee/participatory-guarantee-systems-pgs
(4) The country with the highest organic share in Latin America in 2015 was Uruguay.

Acknowledgements

The authors gratefully acknowledge funding from the Swiss State Secretariat for Economic Affairs (SECO), the International Trade Centre (ITC) and NürnbergMesse, the organisers of the international organic trade fair Biofach.

References and Further Reading

African Union (2011) *Decision on organic farming*. Doc. EX.CL/631 (XVIII). Eighteenth Ordinary Session. 24–28 January 2011, Addis Ababa, Ethiopia. Available at: http://www.au.int/en/sites/default/files/decisions/9646-council_en_24_28_january_2011_executive_council_eighteenth_ordinary_session.pdf (accessed 6 July 2017).

Bouagnimbeck, H. (2011) Organic farming in Africa. In: Willer, H. and Kilcher, L. (eds) *The World of Organic Agriculture. Statistics and Emerging Trends 2011*. FiBL, Frick, Switzerland; IFOAM, Bonn, Germany.

DeMuth, S. (1993) *Community Supported Agriculture (CSA): An Annotated Bibliography and Resource Guide.* Agricultural Research Service, US Department of Agriculture, Beltsville, Maryland. Available at: http://pubs.nal.usda.gov/sites/pubs.nal.usda.gov/files/csadef.html (accessed 6 July 2017).

EOA (2015) *The Ecological Organic Agriculture (EOA)-Initiative. 2015–2025 Strategic Plan*. EOA Continental Steering Committee, African Union Commission, Addis Ababa.

Fair Trade Africa (2014) *Performance Report 2010–2013*. Fair Trade Africa, Nairobi. Available at: http://www.fairtradeafrica.net/wp-content/uploads/2015/01/SOCIAL-IMPACT-REPORT.pdf (accessed 6 July, 2017).

Fair Trade America (2015) *Growing Locally, Connected Globally – Annual Report and Impact Report 2014–2015*. Fair Trade America, Washington, DC. Available at: http://fairtradeamerica.org/~/media/Fairtrade%20America/Files/Reports/Fairtrade_America_Annual_Report_2015_Web2.pdf (accessed 6 July 2017).

Fairtrade International (2015) *Monitoring the Scope and Benefits of Fairtrade*, 7th edn. Fairtrade International, Bonn, Germany. Available at: http://www.fairtrade.net/resources/impact-and-research.html#c8222 (accessed 6 July 2017).

Fairtrade International (2016) *Driving Sales, Deeping Impact. Annual Report 2015–2016*. Fairtrade International, Bonn, Germany. Available at: https://annualreport15-16.fairtrade.net/en/ (accessed 6 July 2017).

FAO (2009) *Glossary on Organic Agriculture*. Food and Agriculture Organization, Rome. Available at: http://www.fao.org/fileadmin/templates/organicag/files/Glossary_on_Organic_Agriculture.pdf (accessed 6 July 2017).

Flores, P. (2014) Organic agriculture in Latin America and the Caribbean. In: Willer, H. and Lernoud, J. (eds) *The World of Organic Agriculture. Statistics and Emerging Trends 2014*. Research Institute of Organic Agriculture (FiBL), Frick, Switzerland; International Federation of Organic Agriculture Movements (IFOAM), Bonn, Germany.

Flores, P. (2015) Organic agriculture in Latin America and the Caribbean. In: Willer, H. and Lernoud, J. (eds) *The World of Organic Agriculture. Statistics and Emerging Trends 2015*. Research Institute of Organic Agriculture (FiBL), Frick, Switzerland; IFOAM – Organics International, Bonn, Germany.

Flores, P. and Soberanes, M. (2016) Organic agriculture in Latin America and the Caribbean. In: Willer, H. and Lernoud, J. (eds) *The World of Organic Agriculture. Statistics and Emerging Trends 2016*. Research Institute of Organic Agriculture (FiBL), Frick, Switzerland; IFOAM – Organics International, Bonn, Germany.

Forum for the Future (2014) *The Future of Tea: a Hero Crop for 2030*. Forum for the Future, London. Available at: http://www.forumforthefuture.org/project/tea0203/ (accessed 4 July 2017).

Gama, J. (2015) Latest developments in organic agriculture in Africa. In: Willer, H. and Lernoud, J. (eds) *The World of Organic Agriculture. Statistics and Emerging Trends 2015*. Research Institute of Organic Agriculture (FiBL), Frick, Switzerland; IFOAM – Organics International, Bonn, Germany.

Hartmann, M., Khalil, S., Bernet, T., Ruhland, F. and Al Ghamdi, A. (2013) Organic agriculture in Saudi Arabia: Country Report 2012. In: Willer, H. and Lernoud, J., Kilcher, L. (eds) *The World of Organic Agriculture. Statistics and Emerging Trends 2013*. Research Institute of Organic Agriculture (FiBL), Frick, Switzerland; International Federation of Organic Agriculture Movements (IFOAM), Bonn, Germany.

Lernoud, J., Potts, J., Sampson, G., Voora, V., Willer, H. and Wozniak, J. (2015) *The State of Sustainable Markets – Statistics and Emerging Trends 2015*. International Trade Centre, Geneva, Switzerland.

Lernoud, J., Potts, J., Sampson, G., Voora, V., Willer, H. and Wozniak, J. (2017) *The State of Sustainable Markets – Statistics and Emerging Trends 2017*. International Trade Centre, Geneva, Switzerland.

Ong, K.W. (2013) Developments in Asia 2012. In: Willer, H., Lernoud, J. and Kilcher, L. (eds) *The World of Organic Agriculture. Statistics and Emerging Trends 2013*. Research Institute of Organic Agriculture (FiBL), Frick, Switzerland; International Federation of Organic Agriculture Movements (IFOAM), Bonn, Germany.

Ong, K.W. (2015) Organic Asia 2014. In: Willer, H. and Lernoud, J. (eds) *The World of Organic Agriculture. Statistics and Emerging Trends 2015*. Research Institute of Organic Agriculture (FiBL), Frick, Switzerland; IFOAM – Organics International, Bonn, Germany.

Ong, K.W. (2016) Organic Asia 2015. In: Willer, H. and Lernoud, J. (eds) *The World of Organic Agriculture. Statistics and Emerging Trends 2016*. Research Institute of Organic Agriculture (FiBL), Frick, Switzerland; IFOAM – Organics International, Bonn, Germany.

Oxfam (2013) *Understanding Wage Issues in the Tea Industry - Report from a Multi-stakeholder Project*. Ethical Tea Partnership with Oxfam International, Oxford, UK. Available at: https://www.oxfam.org/sites/www.oxfam.org/files/oxfam_etp_understanding_wage_issues_in_the_tea_industry.pdf (accessed 4 July 2017).

Panyakul, V. (2016) Thailand: organic market study: methodology and results. In: Willer, H. and Lernoud, J. (eds) *The World of Organic Agriculture. Statistics and Emerging Trends 2016*. Research Institute of Organic Agriculture (FiBL), Frick, Switzerland; IFOAM – Organics International, Bonn, Germany.

Saha, M.K. and Monowar Hossain, S.M. (2014) Organic farming in Bangladesh. In: Willer, H. and Lernoud, J. (eds) *The World of Organic Agriculture. Statistics and Emerging Trends 2014*. Research Institute of Organic Agriculture (FiBL), Frick, Switzerland; International Federation of Organic Agriculture Movements (IFOAM), Bonn, Germany.

Sahota, A. (2016) The global market for organic food & drink. In: Willer, H. and Lernoud, J. (eds) *The World of Organic Agriculture. Statistics and Emerging Trends 2016*. Research Institute of Organic Agriculture (FiBL), Frick, Switzerland; IFOAM – Organics International, Bonn, Germany.

Sahota, A. (2017) The global market for organic food & drink. In: Willer, H. and Lernoud, J. (eds) *The World of Organic Agriculture. Statistics and Emerging Trends 2017*. Research Institute of Organic Agriculture (FiBL), Frick, Switzerland; IFOAM – Organics International, Bonn, Germany.

Willer, H. and Lernoud, J. (eds) (2016) *The World of Organic Agriculture. Statistics and Emerging Trends 2016*. Research Institute of Organic Agriculture (FiBL), Frick, Switzerland; IFOAM – Organics International, Bonn, Germany.

Willer, H. and Lernoud, J. (eds) (2017) *The World of Organic Agriculture. Statistics and Emerging Trends 2017*. Research Institute of Organic Agriculture (FiBL), Frick, Switzerland; IFOAM – Organics International, Bonn, Germany.

Willer, H. and Yussefi, M. (2000) *Ökologischer Landbau weltweit, Statistik und Perspektiven* (Organic farming worldwide, statistics and perspectives). SÖL Special Issue No. 74. Stiftung Ökologie & Landbau (Ecology and Agriculture Foundation), Bad Dürkheim, Germany.

4 Evolution of Producer Organizations in Fair Trade Coffee Certification

Rene T. Capote-Fuentes[1]*, Ulrike Grote[2], Lee Byers[3] and Till Stellmacher[4]
[1]*Coffee, Global Product Management, Fairtrade International, Bonn, Germany;* [2]*Institute for Environmental Economics and World Trade, Leibniz University Hannover, Germany;* [3]*Coffee and Tea, Global Product Management, Fairtrade International, Bonn, Germany;* [4]*Center for Development Research (ZEF), University of Bonn, Germany*

4.1 Introduction

Fair Trade is well known as an international certification system in the context of the global Fair Trade movement. Fairtrade International has the vision of a world in which all small producers and workers can enjoy secure and sustainable livelihoods (Raynolds and Bennett, 2015; Fairtrade International, 2016a). One of its best-known features in product categories such as coffee has to do with providing stable prices to producer organizations of small-scale farmers in developing countries by guaranteeing a minimum price in case the world market price falls below a certain sustainable level. In addition, it provides a price premium that aims to support investments in producers' businesses, livelihoods and communities. The international Fair Trade certification system established its global office in Bonn, Germany, in 1997; it evolved in the 1980s from the initiative of the Dutch Max Havelaar and Mexican coffee growers. The system is now known as Fairtrade International, a shorter version of the legal name of its Bonn global office, Fair Trade Labelling Organizations International (FLO e.V.). The Fair Trade certification system is co-owned by producer- and market-facing stakeholders, namely three producer networks in Latin America, Africa and Asia/Pacific and more than 20 market organizations in countries like Canada, USA, Switzerland, UK and Australia, among others. In recent years, the network of other market organizations has expanded to include South Africa, the Czech Republic, Korea, Hong Kong, Eastern Africa, India, Portugal, Slovakia, Brazil, Taiwan and the Philippines.

Coffee, which is the focus of this chapter, has been a core product in the early and current development of the global Fair Trade certification system in terms of the number of certified producer organizations, sales volume and participating smallholder farmers. In the past two decades, several empirical studies from multiple disciplinary and methodological angles have addressed different types of topics on Fair Trade coffee value chains, but mostly concentrating on farmers rather than producer organizations.

When Fair Trade's development started to mature in the early 2000s (e.g. when the increase in the number of Fair Trade certified producer organizations became a clear trend), Raynolds (2002) and Raynolds *et al.* (2004) were among the first to provide comprehensive papers on producer organizations in both theoretical and

*Corresponding author; e-mail: r.capote@fairtrade.net

empirical dimensions. They hypothesized that successful Fair Trade participation depends on the internal organization and external links of the producer organizations, inter alia. Also, Smith and Loker (2012) and Tellman *et al.* (2011) discussed export activity as a key component of producers' successful participation in Fair Trade. Nevertheless, Reed (2015) highlighted the need for a deeper understanding of the relations between the diverse actors involved in the Fair Trade value chains.

Following this line of argument, the study in this chapter focuses on the characteristics of Fair Trade producer organizations on a global scale and explores their commercial relations with farmers on the one hand and traders on the other hand. Eventually, the producer organizations vastly influence the success of the global Fair Trade movement by increasing the efficiency and the trust among actors, and at the same time reducing the cost and barriers of Fair Trade transactions throughout the value chains. In more detail, the objectives of the study are: (i) to identify patterns of Fair Trade coffee certified producer organizations; and (ii) to analyse the evolution of these organizations in Fair Trade coffee.

This chapter is unique in its approach to present an exploratory study on a group of 70 producer organizations that had been Fair Trade coffee certified for 10 years or longer (as of the end of 2011). This group of 70 producer organizations can be considered as long-term organizations in the global group of 348 producer organizations holding Fair Trade coffee certification by the end of 2011 worldwide. In addition, this dataset is supplemented by more general data on all existing Fair Trade coffee producer organizations between 2011 and 2014. The chapter briefly reviews the literature on Fair Trade coffee in Section 4.2. A description of the data follows in Section 4.3, along with an outline of the methodologies applied. Section 4.4 presents the results and Section 4.5 discusses, summarizes and concludes.

4.2 Literature Review and Conceptual Framework

There are different strands in the literature. A number of papers look at the overall governance of coffee value chains. Producer organizations do play a major role within these value chains since they are the ones to be certified on the producer side. While only a few papers focus on the forward linkages (Raynolds, 2002, 2009; Utting, 2009; Smith and Loker, 2012), many papers analyse the backward linkages, assessing the impacts of certification on small-scale producers (Raynolds *et al.*, 2004; Arnould *et al.*, 2009; Utting, 2009; Beuchelt and Zeller, 2011; Tellman *et al.*, 2011; Elder *et al.*, 2012; Fitzgerald, 2012).

Raynolds (2002) departed from the observation that the Fair Trade network aims to establish more egalitarian value chains. She concluded that Fair Trade indeed created new networks and interlinkages between actors with a decreasing social distance between producers and consumers. Utting (2009) presented an impact assessment framework that allowed the evaluation of local-level impacts resulting from 'responsible trade' interventions such as Fair Trade and ethical trade. The study applied the framework to a coffee-producing region in northern Nicaragua and assessed the changes (or impacts) promoted by a Fair Trade intervention with regard to livelihood impacts, organizational impacts and institutional impacts. In a specific producer organization, Utting (2009) discussed how secondary and tertiary associations help primary-level cooperatives to sell their coffee directly to buyers in developed countries, bypassing the need for intermediaries in the value chain.

Smith and Loker (2012) explored the experiences of a cooperative of small to medium coffee growers in rural Honduras. In more detail, they examined the factors responsible for its ability to sell coffee on the international market through organic and Fair Trade channels, as well as the market entry barriers it faced (resource barriers, communication/information barriers, organizational barriers, knowledge barriers and administrative barriers). They also analysed characteristics of the cooperative such as: (i) organizational and legal setup; (ii) number of affiliated members; (iii) average size of farms; (iv) to what extent members organize production and post-harvesting processing individually or rely on the cooperative; (v) capacity to acquire processing infrastructure; (vi) production volume and extension of land; (vii) holding of organic certification and capacity to ensure quality of coffee at all stages of production and processing; (viii) fluctuations in the number of members; (ix) communication infrastructure; and (x) administrative capacity.

Raynolds (2009) discussed mainstreaming in Fair Trade coffee. When analysing the producers' side of the impacts of diverse buyer relations on producer groups, she clarified the range of producer organizations the study built on. These cooperatives embodied key variations among Fair Trade coffee suppliers in terms of age (years being Fair Trade certified), number of members, annual export volumes, substantial portion of export volume as Fair Trade sales due to high quality of coffee fitting market demand, and number of Fair Trade buyers from three to six buyers.

With respect to the backward linkages, Elder *et al.* (2012) assessed the impact of Fair Trade on social capital – measured as farmer trust and participation – amongst Rwandan coffee farmers. They specifically focused on understanding whether changes in producer-level social capital were primarily a result of cooperative organization or of Fair Trade certification. They found a negative association between Fair Trade and farmer trust and a positive association with a perceived higher level of participation of women. Fitzgerald (2012) examined to what extent Fair Trade promoted community cohesion and coordination as Fair Trade coffee standards required farmers to be affiliated to a producer organization. He elaborated on organizational issues, such as cooperatives: (i) deepening their involvement with the coffee trade to extend vertically up the commodity chain; (ii) strong cooperatives being able to construct processing facilities and expanding into other niche markets such as organic; (iii) enabling producers to pool resources and benefiting from economies of scale; (iv) connecting with other Fair Trade participating cooperatives thereby creating secondary-level cooperatives that could augment advocacy strength; and (v) the role cooperatives played in administering the social premium towards collective purposes.

Tellman *et al.* (2011) studied the practical obstacles to participation in Fair Trade in El Salvador by comparing two coffee cooperatives from Fair Trade value chains. They identified different barriers to Fair Trade certification for the two cooperatives related to certification costs, economies of scale, lack of government support for cooperatives, corruption and quality constraints. Farmers who received higher coffee prices via exports were not necessarily better off than their non-exporting counterparts; their actual revenues were also affected by issues such as costs and debts of the producer organization. Raynolds *et al.* (2004) compared the experiences of seven Fair Trade coffee producer organizations in Mexico, Guatemala and El Salvador. They stressed the importance of these producer organizations in creating and maintaining strong external ties with buyers, which in turn reflected on volume of exports and volume of Fair Trade exports. They found that producer organizations and their members derived important benefits from Fair Trade. They stressed the importance of financial benefits of Fair Trade in the short run while capacity building was expected to be the most important aspect in the long run.

Many studies assessed the welfare impact of Fair Trade certification at household and cooperative level. Ruben and Fort (2011) did a rigorous quantitative impact assessment of Fair Trade certification. They found only modest direct income and production effects, but significant changes in organization, input use, wealth and assets, and risk attitudes. Jena *et al.* (2012) also found that the certification of coffee cooperatives had in total only a low impact on small-scale coffee producers' livelihoods, mainly due to low productivity, an insignificant price premium and poor access to credit and information from the cooperative. They stressed the large differences in production and organizational capacities between the local cooperatives and that 'good' cooperatives had reaped the benefits of certification.

Beuchelt and Zeller (2011) quantified the impact of randomly selected farmers of conventional, organic and double-certified organic–Fair Trade producer organizations on poverty and income. They found that farm-gate prices of certified coffees were higher than of conventional coffees, but certified producers were more often found below the absolute poverty line than conventional ones. Over a period of 10 years, organic and organic–Fair Trade farmers became even poorer relative to conventional producers. Chiputwa *et al.* (2015) analysed the impact of Fair Trade, organic and UTZ (sustainable farming) certification on the livelihoods of small-scale coffee producers in Uganda. They found that Fair Trade certification increased household living standards by 30% and significantly reduced poverty. However, for the other two certification schemes, they did not find any significant impacts.

Arnould *et al.* (2009) also focused on the grower level and analysed whether participation in a Fair Trade coffee marketing channel delivered benefits to small-scale producers in Latin America. The authors recognized that differences in the size of Fair Trade cooperatives (i.e. number of members) might have an impact, such that larger cooperatives were expected to offer better services to their members. The authors used a stratified cluster sampling in order to control threats to the validity of Fair Trade versus non-Fair Trade comparisons. The Fair Trade cooperatives in a country were stratified into three groups (small, medium and large) according to the number of members of the cooperative. However, contrary to the authors' a priori expectations, the analyses revealed no consistent patterns of difference based on cooperative or community size and so the authors decided not to report those results.

In sum, most empirical studies concentrated at the farm level. The producer organizations – as the crucial entities in which farmers can organize themselves and which eventually hold the Fair Trade certification – have not been considered as widely (Ruben, 2008; Arnould *et al.*, 2009; Utting, 2009; Elder *et al.*, 2012; Fitzgerald, 2012; Raynolds and Bennett, 2015). Furthermore, these studies at the farm level often ignored which types of umbrella producer organizations the farmers were affiliated to. Empirical studies on Fair Trade coffee certified producer organizations have also typically not been global in scope. They generally addressed only one or a few organizations in a country or a region – a number that is substantially lower than the total number of worldwide certified organizations in the specific year (Raynolds *et al.*, 2004; Barham and Weber, 2012; Raynolds and Bennett, 2015).

Against the background of the literature, the producer organizations are at the centre of our focus as little is known about them ('black box'). We have framed the analysis of the evolution/integration of producer organizations in Fair Trade coffee in the two main value chains that exist in Fair Trade coffee: in the first one, the producer organizations directly sell to exporters; and in the second one, the producer organizations sell to importers or exporters. Producer organizations are the stakeholders, who on the production side of the product chain hold the Fair Trade certification. However, they do not always export directly, i.e. they do not always maintain commercial relations with importers but with exporters.

In the Fair Trade standards related to coffee, attention is directed to the export operation of producer organizations, or at least of the possibility of it. The Fair Trade Standards for Small-Producer Organizations (Fairtrade International, 2016b), the Trader Standard (Fairtrade International, 2016c) and their corresponding compliance criteria do not necessarily assume that producer organizations must have direct commercial relations with importers. However, in the Coffee Standard (Fairtrade International, 2016d), which is specific to coffee, and some of the general topics outlined in the Standards for Small-Producer Organizations and Traders, a focus is directed to the export level. For example, the Fair Trade coffee price to be paid to a producer organization (either the reference market price or the minimum price, whichever is higher) is set at the Free on Board (FOB) level with reference to the original producer country.

4.3 Data and Methodology

We use unique data, originating from Fairtrade International, on the producer organizations and the members (farmers) affiliated to them. The data consist of two sets that capture different time periods and that differ in complexity. The first dataset relates to the period 2011–2014, providing some general information on all existing Fair Trade coffee producer organizations. The second set is a sample of 70 Fair Trade coffee certified producer organizations, based on all 348 Fair Trade producer organizations that existed in 2011. This sample includes nearly all coffee producer organizations that had been certified for 10 years or longer as of the end of 2011. Within the sample, the figures were then compared with the reference year 2001. The information available derives from the obligatory and regular reports of the Fair Trade producer organizations to the Fair Trade system. These reports have been used to create this unique dataset, always observing confidentiality guidelines. The general information provided by the producer organizations relates, for example, to their area under Fair Trade cultivation, production volume and sales volumes (total, export and Fair Trade). In addition, their size, country of operation and some more information on their individual members (farmers) are provided.

This exploratory study conducted principal component and cluster analyses to identify the variables most useful in distinguishing patterns of producer organizations and to classify the organizations more systematically. The continuous variables were converted into categorical ones and a principal components analysis for categorical data was run. The cluster analysis was conducted through a two-step procedure (distance measure: log-likelihood). The cluster analysis included both categorical and continuous variables (respectively, 12 and 10 variables). Continuous variables were automatically standardized by the two-step cluster analysis procedure. No specific number of clusters was pre-decided, i.e. it was left for the algorithm to indicate the resulting number of clusters. Both the Schwarz Bayesian criterion and the Akaike information criterion resulted in three clusters. No outlier cases (producer organizations) were indicated by the cluster analysis at 25% of noise handling. Thereafter, we analysed the evolution of these organizations in Fair Trade. We did this by examining, between 2001 and 2011, the change in producers' sales volumes, for instance, and whether producer organizations maintained commercial relations with importers, or only with exporters.

4.4 Results on the Fair Trade Producer Organizations

4.4.1 Overall characteristics

By the end of 2011, 348 producer organizations were holding a Fair Trade coffee certification. In 2014, a total of 445 producer organizations were already certified for Fair Trade coffee. The number along with the production volume of Green Bean Equivalent (GBE) has thus constantly increased over the years (Table 4.1). In addition, it is worth mentioning that during the period 2011–2014, around 35% of the global Fair Trade certified production volume was also organic certified (Table 4.1).

Some interesting characteristics of the producer organizations derive from the sample of 70 Fair Trade coffee certified producer organizations. We call these the pioneer organizations, since they are the producer organizations that had been certified for 10 years or longer by the end of 2011. Pioneer organizations had existed for at least a few years before becoming Fair Trade coffee certified. Their organizational existence spanned over 12 years, and 14 out of the 70 organizations had existed for at least 32 years. Pioneer organizations had held a Fair Trade certification for at least 10 years, and 10 out of 70 had been Fair Trade certified for 19–22 years. They focused on Fair Trade washed Arabica coffee; only four out of 70 held a Fair Trade certification for a product other than coffee; and 11 out of 70 produced or traded products other than coffee outside of the Fair Trade certification.

Most of these Fair Trade coffee certified producer organizations (67 out of 70) are based in Latin America and the Caribbean, i.e. the region where Fair Trade certification was first introduced in the 1990s. The number of organizations per country in the sample was: Bolivia, 8; Brazil, 1; Colombia, 10; Costa Rica, 1; Dominican Republic, 1; El Salvador, 1; Guatemala, 4; Honduras, 11; Mexico, 19; Nicaragua, 3; Peru, 8; Papua New Guinea, 1; and Tanzania, 2.

The size of the producer organizations differs largely; it ranges between 39 and 58,126 members (farmers). The median amounts to 392 members. With respect to gender, the median amounts to 83 female farmers being members; the minimum and maximum reach 3 and 4200 female members, respectively. Although the producer organizations differ in their legal form, nearly half (30 out of 70) are cooperatives. Producer organizations typically have two main structures recognized as first- and second-grade organizations. Nineteen out of the 70 organizations were second-grade. In first-grade organizations, individual members are directly affiliated to the organization, whereas a second-grade organization includes a number of first-grade organizations with their individual members. Second-grade organizations are typically formed to bring together first-grade organizations over an extended territory. The second level in the structure is often dedicated to external commercial and sales operations beyond the organization, while its first-grade levels deal with sourcing the coffee from individual members.

Table 4.1. Number of Fair Trade coffee certified producer organizations and volumes of Fair Trade and organic certified coffee worldwide (based on Fairtrade International, 2015).

Year	No. of producer organizations	Total production volume (t, GBE)[a]	Of which organic certified (t, GBE)[a]	Percentage organic certified vs. total volume
2014	445	549,400	185,300	34
2013	439	473,600	176,500	37
2012	402	398,800	149,300	37
2011	348	393,000	134,400	34

[a]t, metric ton (1000 kg); GBE, Green Bean Equivalent.

4.4.2 Activities and certification status

Regarding coffee volumes, the mean production of the organizations was 1106 t GBE. Sales (total, export and Fair Trade) amounted to 875, 798 and 562 t GBE, respectively. The individual members (farmers) deliver coffee to their corresponding producer organization in the form of: (i) only cherry (i.e. non-processed coffee fruit); (ii) cherry or parchment (i.e. endocarp); or (iii) only parchment. All in all, 44 producer organizations reported to have received parchment from their farmers; another 13 received either cherry or parchment, and another 13 only unprocessed cherry.

The organizations then typically conduct further post-harvest processing of the coffee delivered to them by farmers and deliver parchment or green coffee to the clients. Out of the 70 organizations, 14 indicated to have delivered parchment to their clients and 56 delivered green coffee, which is typically coffee ready for export.

Owning post-harvest processing facilities is an important feature of the producer organizations, as it shapes the trade relation of the organizations with its members and the wider coffee sector. All in all, 47 out of the 70 producer organizations had their own post-harvest processing installations. Regarding the post-harvest steps that these installations enable in a typical washed Arabica process, seven organizations had wet mills, 28 organizations had installations of the next processing step (e.g. dry mills) and 12 organizations owned installations that enabled further processing of the coffee to be exported in the typical green bean form (e.g. installations for hulling, grading and export preparation) or even roasting of the coffee.

Many producer organizations (56 out of 70) indicated to have a focus on coffee quality as reflected in their organizational mission with, for example, giving priority to the specialty (high-quality) segment of the coffee market, having regular investments to continuously develop their quality professionals and upgrade their coffee-cupping laboratories, and having obtained awards in national or international quality competitions.

A total of 66 of the 70 pioneer organizations held an organic certification by 2011 and 63% of their total coffee production was both Fair Trade and organic certified, i.e. around 48,842 out of 77,456 t GBE. In addition, although the pioneer organizations focused on Fair Trade coffee, 18 out of 70 held a certification other than Fair Trade or organic (e.g. UTZ or Rainforest Alliance certifications) or were involved in a verification programme (e.g. 4C, Nespresso's AAA, Starbucks's Café Practices).

The organizations also differ to a large extent depending on the number of Fair Trade clients: 34 producer organizations indicated that they had between one and three clients, another 22 had four to eight clients and 14 had up to 27 Fair Trade clients. In addition, 53 out of 70 were having commercial relations with importers, while 17 had only exporters as their commercial clients.

4.4.3 Producer organizations' clusters

As mentioned in Section 4.3, the data that enabled the classification by cluster analysis consist of a sample of 70 Fair Trade coffee certified producer organizations, out of the 348 Fair Trade producer organizations that existed in 2011. This 70-organization sample included nearly all coffee producer organizations that had been certified for 10 years or longer as of the end of 2011.

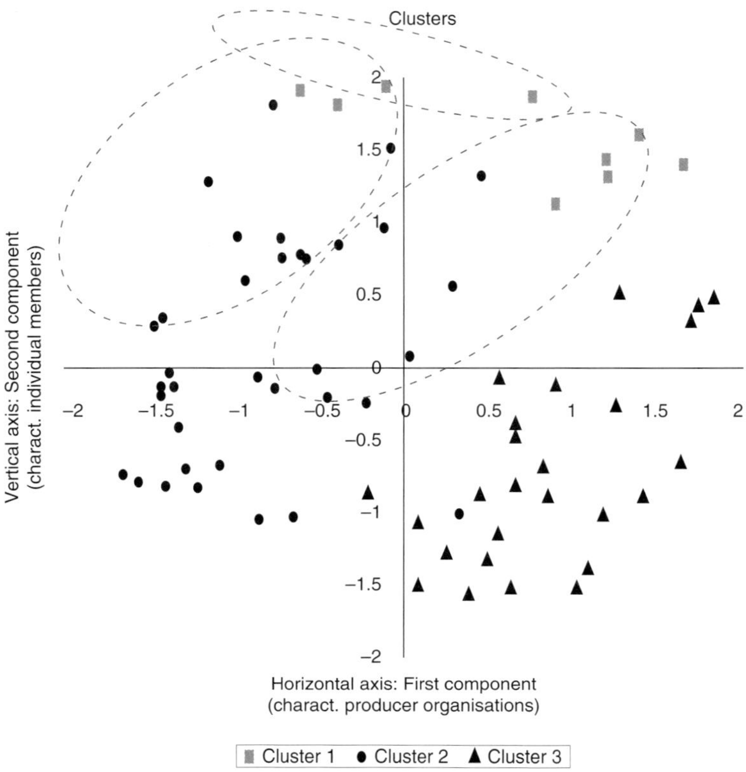

Fig. 4.1. Three clusters of pioneer Fair Trade coffee certified producer organizations.

The classification analysis of the 70 producer organizations yielded three clusters with nine, 35 and 26 organizations (Fig. 4.1). Three components accounted for 70% of the total variance (39%, 24% and 7%, respectively). The first component (represented in the horizontal axis) mainly relates to variables that characterize the producer organizations, while the second component mainly relates to variables that characterize the individual members.

Component 1 on the characteristics of the producer organizations tends to have somewhat bigger loadings from variables like number of individual members, area under Fair Trade cultivation, production volume, total sales volumes, exports and Fair Trade. Component 2 on the characteristics of the individual members tends to have somewhat bigger loadings from variables on the individual members' volumes, namely of production and total sales, exports and Fair Trade.

The characteristics of the three clusters can be summarized with some of the variables that were significant to define the clusters.

Cluster 1 (Table 4.2) groups producer organizations with big volumes (mean production volume 2009 t) with small to average number of members (mean 325). Individual members have around average farm size (mean 5.07 ha) but above-average production volume (mean 5.52 t) and Fair Trade sales volume (mean 3.98 t). This cluster mainly groups producer organizations having commercial relations with importers (Table 4.2). These organizations are all cooperatives and all focus on quality. Although this cluster mainly groups producer organizations having commercial relations with importers, none of these organizations owns post-harvest processing installations of the kind that would enable preparing coffee for export.

Cluster 2 (Table 4.2) groups producer organizations with small volumes (mean production

Table 4.2. Clusters of producer organizations (POs).

Cluster	POs: no. (%)	Production volume (t)	No. of members (mean)	Farm size individual members (ha, mean)	Average production volume (t)	Fair Trade sales volume (t)
1	9 (13%)	2009	325	5.07	5.52	3.98
2	35 (50%)	421	242	6.02	1.66	0.65
3	26 (37%)	1715	4828	3.47	0.77	0.40
All cases combined (when not clustered)	70 (100%)	1106	325	4.95	1.83	0.98

volume 421 t) with small to average number of members (mean 242). Individual members have around average farm size (mean 6.02 ha) and production volume (mean 1.66 t) but below-average Fair Trade sales volume (mean 0.65 t). This cluster is a more mixed group of organizations that have commercial relations with either importers or exporters. The coffee from these producer organizations delivered to their clients, when compared with clusters 1 and 2, has less predominance of green coffee, which is the form typically delivered to exporters. Organizations in cluster 2 tend to have the smallest number of Fair Trade clients.

Cluster 3 (Table 4.2) groups producer organizations with above-average volumes (mean production volume 1715 t) and above-average number of members (mean 4828). Individual members have below-average farm size (mean 3.47 ha), production volume (mean 0.77 t) and Fair Trade sales volume (0.40 t). This cluster, similarly to cluster 1, mainly includes producer organizations having commercial relations with importers. This cluster tends to have more second-grade organizations. At the same time this cluster's organizations do not own coffee post-harvest processing installations of the kind that would enable the first processing steps of the coffee fruits once they have been picked from the trees.

4.4.4 Evolution of producer organizations

When looking at the evolution of pioneer producer organizations in Fair Trade coffee between 2001 and 2011 (Table 4.3), the organizations became somewhat smaller with regard to the median number of individual members (from 440 to 392). At the same time, the organizations achieved somewhat bigger median volumes of production (from 317 to 382 t), total sales (changed from 203 to 309 t) and export sales (changed from 154 to 254 t). Mean values did not seem to change as much (Table 4.3).

The number of organizations that were having direct commercial relations with importers changed from 33 to 53 out of 70 from 2001 to 2011. Compared with the producer organizations, individual members changed less in median values and more in mean values. Between 2001 and 2011, individual members achieved somewhat bigger mean volumes of production (from 1.16 t to 1.83 t), total sales (from 0.77 t to 1.23 t) and export sales (from 0.61 t to 1.10 t).

Figure 4.2a further indicates some of the changes that took place at organizations which, from 2001 to 2011, changed from having commercial relations with exporters into starting to work with importers, or vice versa. Out of the 70 organizations, 17 were selling to exporters both in 2001 and 2011; 17 were selling to exporters in 2001 but then were selling to importers in 2011; and 36 organizations were selling to importers both in 2001 and 2011.

Organizations (and their members) that, from 2001 to 2011, changed from selling to exporters into selling to importers had the most distinctive changes in median volumes (Fig. 4.2a, group of organizations named as E-I). This was particularly noticeable in the median volumes of individual members (Fig. 4.2b, group E-I). Boxplots of organizations that worked with exporters in both 2001 and 2011 (Fig. 4.2a, group E-E) indicate the following changes in the

Table 4.3. Evolution of producer organizations in Fair Trade coffee between 2001 and 2011.

		Producer organizations		Individual members	
		2001	2011	2001	2011
No. of individual members	Median	440	392	*	
	Mean	2093	1957	*	
Total production volume (t)	Median	317	382	0.80	0.81
	Mean	1077	1107	1.16	1.83
Total sales volume (t)	Median	203	309	0.58	0.62
	Mean	719	876	0.77	1.23
Export sales volume (t)	Median	154	254	0.39	0.54
	Mean	500	798	0.61	1.10
Fair Trade sales volume	Median	**	237	**	0.47
	Mean	**	562	**	0.99
Direct commercial relations	With importers (no. of organizations)	33	53	***	
	With exporters (no. of organizations)	37	17	***	

*, ***Variables do not apply to individual members but to organizations; **2001 Fair Trade sales data were not consistently available across all organizations. As indicative reference, 2001 Fair Trade sales volume was typically lower than (or in some cases equal to) the 2001 export volume.

organizations' median volumes and distribution. Median production volumes were similar in both years; however, in 2011 the upper range (above median value) became even bigger than it already was in 2001. The median total sales volume tended to be somewhat smaller in 2011; the upper range (above median value) became somewhat more spread. Median export sales volumes were similar in both years, while the upper and lower range around the median became somewhat more spread.

Boxplots of organizations that were selling to exporters in 2001 but then changed into selling to importers in 2011 (Fig. 4.2a, group E-I) indicate the following changes in the organizations' median volumes and distribution. Median production volumes in 2011 became somewhat bigger; and both the upper range (above median value) and lower range (below median value) became somewhat more spread. The total sales volume and export sales volume had similar changes: median volumes tended to be somewhat bigger in 2011; the upper range (above median value) and lower range (below median value) became somewhat more spread in both variables.

Boxplots of organizations that worked with importers in both 2001 and 2011 (Fig. 4.2a, group I-I) indicate the following changes in the organizations' median volumes and distribution. Production volumes, in regard to both the median and the ranges around the median, were rather similar in 2001 and 2011. The median total sales volume became somewhat smaller; and the lower range (below median value) became somewhat less spread. Median export sales volume was bigger in 2011; the upper range (above median value) became more spread and the lower range (below median value) remained rather similar.

Boxplots of individual members of organizations that worked with exporters in both 2001 and 2011 (Fig. 4.2b, group E-E) indicate the following changes in the members' median volumes and distribution. Median production volumes were similar in both years, while in 2011 the upper range (above median value) became bigger than it was in 2001. The median total sales volume remained similar in 2011; the lower range (below median value) became somewhat more spread. The median export sales volume became somewhat higher in 2011; the upper range (above median value) shrank.

Boxplots of individual members of organizations that were selling to exporters in 2001 but then had changed into selling to importers in 2011 (Fig. 4.2b, group E-I) indicate the following changes in the members' median volumes

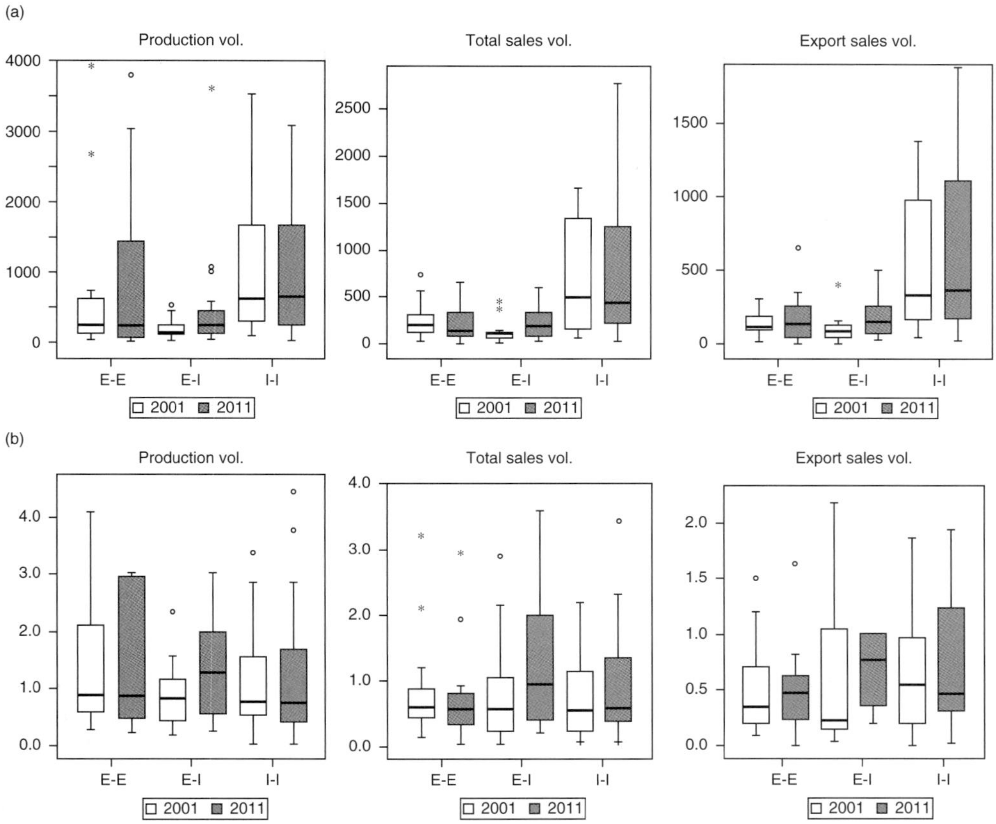

Fig. 4.2. Three clusters of pioneer Fair Trade coffee certified producer organizations: (a) production and sales volumes of the producer organisations (t, GBE); (b) production and sales volumes of the individual members of the organizations (t, GBE). Three graphs, from left to right: production volume; total sales volume; and export sales volume. Each graph includes a set of two boxplots for each of the three following groups of producer organizations (as named in the horizontal axis of each graph): (E-E): producer organizations selling to exporters both in 2001 and 2011 (n = 17); (E-I): producer organizations selling to exporters in 2001 and to importers in 2011 (n = 17); (I-I): producer organizations selling to importers both in 2001 and 2011 (n = 36).

and distribution. Median production volumes in 2011 became somewhat bigger; and both the upper range (above median value) and lower range (below median value) became somewhat more spread. Median of total sales volume and export sales volume became somewhat bigger in 2011. The upper range (above median value) of the total sales volume became somewhat more spread, while that of export sales volume shrank.

Boxplots of individual members of organizations that worked with importers in both 2001 and 2011 (Fig. 4.2b, group I-I) indicate the following changes in the members' median volumes and distribution. Median production volume remained similar; the lower and upper ranges became somewhat more spread. Median total sales volume remained similar, while that of export sales volume became smaller; their upper ranges became more spread and the lower ranges shrank.

4.5 Discussion and Conclusions

This exploratory study aimed to (i) identify patterns of Fair Trade coffee certified producer organizations and (ii) analyse the evolution of these organizations in Fair Trade coffee. The analysis has shown that there is a large diversity of Fair Trade certified producer organizations and individual members. Although the Fair Trade standards are a single reference that applies worldwide, the organizations and their members are not a homogeneous group. Organizational functioning, trade logistics and relations between the organizations and their individual members can substantially differ between organizations that range from a few dozens or hundreds of members to organizations that have thousands of members. This diversity of organizations and members can bring differences in the development paths that producer organizations and their members would be pursuing with the complementary enabling contribution of Fair Trade. A range of questions arise when producer organizations and their members discuss their baseline and target values. For example: (i) what a reasonable minimum and maximum number of members would be; (ii) whether there is a minimum volume or value of trade below which costs cannot be covered; and (iii) how continental or country-specific contexts shape these questions and answers.

The coexistence of Fair Trade and organic coffee continues to play an important role for producer organizations. Although currently there are Fair Trade certified producer organizations in around 30 countries, the relevance of the organic segment has been a landmark since Fair Trade's beginnings because, on the production side, the certification started in Latin American regions with a tradition in organic production, for example in Mexico.

This double certification of organizations is further enabled and supported by trends in sustainability requirements and expectations in some consumer countries. An important element supporting the existence of the double certification on the producers' side is the price incentive that Fair Trade brings to organic certified coffee. Besides the Fair Trade premium to producer organizations of \$0.20/lb (\$0.44/kg) of green coffee, Fair Trade coffee certification enforces a price differential of \$0.30 when the product is also organic certified. Another element of context is that, at the farm level (i.e. at the level of individual members of producer organizations), Fair Trade focuses on the small-scale growers' production and not on the larger plantations. Small-scale growers tend to have rather traditional production practices in coffee plots that are often in shade areas. This frequently provides the basis for organic practices and certification, which is then taken up at the level of producer organizations when these are formed and certified.

Pioneer producer organizations have undergone important changes towards concentration in a decade. They have become smaller in median numbers of individual members and, at the same time, they have reached bigger volumes of coffee. Individual members have mainly changed in mean values but not so much in median values. Some decrease in number of individual members can be partially explained by a natural process that can take place at a certified organization. The status of the membership is the key information gathered in Fair Trade audits as it provides, for example, evidence on topics like democratic decisions on the use of the Fair Trade premium payments and traceability of the coffee. Thus, organizations tend to increase the accuracy of reporting on membership, often leading to some decrease in the number of members as those who do not actively participate in key assemblies and deliver coffee would drop their membership. Also, from larger organizations, smaller groups of members evolve and often create new smaller organizations that achieve their own Fair Trade certification.

Changes in sales and export volumes at the organizations probably reflect the effect of certification as co-driver of changes in trade at the organizational level. Because the Fair Trade certificate is held at the organizational level, the new level of operation and strategy that the certification brings beyond members has a strong focus on the organizational level. With regard to the individual members, though more research would be needed, the effect on their median volumes is not as visible across all organizations. Changes in median volumes of individual members were most evident in organizations that changed from selling to exporters into selling to importers.

Having commercial relations with importers also coincided with a somewhat higher number of clients. However, a word of caution is needed before recommending the development of commercial relations with importers or exporters. The scope of this study has not included other relevant topics like an analysis of costs and benefits and thus such a recommendation would go beyond the data evidence of this study. Nonetheless, some light can be shed with regard to the ownership of processing installations by producer organizations. Reed (2015) indicated that a more nuanced analysis of the complex nature of Fair Trade, especially the relationships between its key economic actors, could provide better insights into functioning and its potential for promoting development.

Achieving relationships with importers has at least been indirectly influenced by the Fair Trade standards. An important element in the Fair Trade standards is that the Fair Trade reference price (and the Fair Trade minimum price, FMP) to be paid to producer organizations is defined at the export level (FOB) of the producer country. A complementary element inducing producer organizations to consider exporting by themselves is striving for empowerment in the Fair Trade discourse, which on the producers' side would manifest as the wish to deal less with local traders and to seek direct relations with clients in importer countries.

However, fully owning post-harvest processing installations for dry-processing and export preparation is not always a pre-condition to being able to deliver green coffee to importers. Organizations that did not always own processing installations for export preparation (in cluster 1) were typically delivering green coffee ready for export to their clients. These organizations were even receiving from their individual members a coffee form that had undergone less post-harvest processing; therefore these organizations solved the further processing of coffee yet without owning processing installations for export preparation.

Producer organizations should be careful when considering the acquisition of export processing installations. The organization might find itself not having volumes of production high enough in order to control the costs of the export processing installations. If owning post-harvest processing installations is not a pre-condition to achieve export capacity, then producer organizations could consider opportunities for export collaboration with other organizations, outsourcing export services, or partially owning processing facilities. Cases for collaboration that are found in Fair Trade coffee include the following.

- Producer organizations jointly set up and own an exporter enterprise.
- Producer organizations export through another producer organization that has become also certified as an exporter trader. A trader status allows this second organization to export not only the produce of its member coffee growers but also the produce of other organizations.
- Producer organizations, while continuing to work with one or different exporters, increasingly develop relations with importers and other market players in importer countries. In this way producer organizations aim to play a stronger role in co-driving the commercial relation with the exporter while the exporter is accountable for the export logistics and infrastructure.

When producer organizations shape their trade context and assess whether to develop commercial relations with exporters or importers, not only the economic and cost/benefits elements will be of importance. The possibility of increasing the autonomy and empowerment of the producer organization and those of its individual members would typically play an important role. The development of Fair Trade networks of producers at country, regional and global levels can continue to provide a framework where producers can further assess and develop collaboration schemes in trade and other topics.

References

Arnould, J.A., Plastina, A. and Ball, D. (2009) Does fair trade deliver on its core value proposition? Effects on income, educational attainment, and health in three countries. *Journal of Public Policy & Marketing* 28(2), 186–201.

Barham, B.L. and Weber, J.G. (2012) The economic sustainability of certified coffee: recent evidence from Mexico and Peru. *World Development* 40(6), 1269–1279.

Beuchelt, T.D. and Zeller, M. (2011) Profits and poverty: certification's troubled link for Nicaragua's organic and fair trade coffee producers. *Ecological Economics* 70(2011), 1316–1324.

Chiputwa, B., Spielman, D. and Qaim, M. (2015) Food standards, certification, and poverty among coffee farmers in Uganda. *World Development* 66, 400–412.

Elder, S.D., Zerriffi, H. and Billon, P.L. (2012) Effects of fair trade certification on social capital: the case of Rwandan coffee producers. *World Development* 40(11), 2355–2367.

Fairtrade International (2015) Fair Trade Fact file: Coffee. In: *Monitoring the Scope and Benefits of Fair Trade*, 7th edn. Fairtrade International, Bonn, Germany, pp. 74–79. Available at: http://www.fairtrade.net/impact-and-research.html (accessed 5 February 2017).

Fairtrade International (2016a) *Changing Trade, Changing Lives, 2016–2020. Fair Trade Global Strategy*. Fairtrade International, Bonn, Germany. Available at: http://www.fairtrade.net/ (accessed 5 February 2017).

Fairtrade International (2016b) *Fair Trade Standards for Small Producer Organizations*. Current version: 01.05.2011 v1.3. Fairtrade International, Bonn, Germany.

Fairtrade International (2016c) *Fair Trade Trader Standard*. Current version: 01.03.2015 v1.1. Fairtrade International, Bonn, Germany.

Fairtrade International (2016d) *Fair Trade Standard for Coffee for Small Producer Organizations and Traders*. Current version: 01.04.2011 v1.3. Fairtrade International, Bonn, Germany.

Fitzgerald, C.S. (2012) Fair Trade as a community development initiative: local and global implications. *Advances in Social Work* 13(2), 375–390.

Jena, P.R., Chichaibelu, B.B., Stellmacher, T. and Grote, U. (2012) The impact of coffee certification on small-scale producers' livelihoods: a case study from the Jimma Zone, Ethiopia. *Agricultural Economics* 43(4), 429–440.

Raynolds, L.T. (2002) Consumer/producer links in fair trade coffee networks. *Sociologia Ruralis* 42(4), 404–424.

Raynolds, L.T. (2009) Mainstreaming fair trade coffee: from partnership to traceability. *World Development* 37(6), 1083–1093.

Raynolds, L.T. and Bennett, E.A. (eds) (2015) *Handbook of Research on Fair Trade*. Edward Elgar Publishing, Cheltenham, UK, and Northampton, Massachusetts.

Raynolds, L.T., Murray, D. and Taylor, P.L. (2004) Fair trade coffee: building producer capacity via global networks. *Journal of International Development* 16(8), 1109–1121.

Reed, D. (2015) Cooperatives, corporations and fair trade. In: Raynolds, L.T. and Bennet, E.A. (eds) *Handbook of Research on Fair Trade*. Edward Elgar Publishing, Cheltenham, UK, and Northampton, Massachusetts, pp. 211–229.

Ruben, R. (2008) The development impact of fair trade: from discourse to data. In: Ruben, R. (ed.) *The Impact of Fair Trade*. Wageningen Academic Publishers, Wageningen, The Netherlands, pp.19–47.

Ruben, R. and Fort, R. (2011) The impact of fair trade certification for coffee farmers in Peru. *World Development* 40(3), 570–582.

Smith, E. and Loker, W.M. (2012) 'We Know Our Worth': lessons from a fair trade coffee cooperative in Honduras. *Human Organization* 71(1), 87–98.

Tellman, B., Gray, L.C. and Bacon, C.M. (2011) Not fair enough: historic and institutional barriers to fair trade coffee in El Salvador. *Journal of Latin American Geography* 10(2), 107–127.

Utting, K. (2009) Assessing the impact of fair trade coffee: towards an integrative framework. *Journal of Business Ethics* 86, 127–149. doi: 10.1007/s10551-008-9761-9.

5 Methodological Toolbox

Hermann Waibel and Priyanka Parvathi*
Institute of Development and Agricultural Economics, Leibniz University Hannover, Germany

5.1 Introduction

Rigorous scientific studies on organic and Fair Trade agriculture can be undertaken with standard economic methodologies such as adoption studies, impact assessment, cost–benefit analysis and environmental economic analysis. However, organic and Fair Trade agriculture is expected to have effects that cannot be fully captured by standard economic analysis that is focused on economic efficiency criteria. Organic products address the health and environmental concerns of consumers and therefore environmental sustainability and health economics should be incorporated in the analysis. Fair Trade incorporates a pro-poor social premium in the price of their products and so distributional aspects with regards to wealth and access to resources should be included in the study. Methods that go beyond neoclassic welfare analysis, like assessment of indicators for ecological and social sustainability, poverty reduction and long-term wealth effects, are necessary tools that complement economic analysis in order to arrive at more meaningful conclusions.

In the following, the main methodological tools for the conduct of research for organic agriculture and Fair Trade are introduced. Two groups of tool boxes are presented. Firstly, we review methods that are based on the tenets of welfare economics and can be categorized under the umbrella of cost–benefit analysis. Secondly, we discuss approaches that are complementary to the standard economic approaches and could be labelled as mixed quantitative–qualitative analysis.

5.2 General Framework

Although organic agriculture and Fair Trade projects are often initiated by private agents, their economic assessment must be undertaken from a societal point of view. This is because their motivation stems from the market failures of conventional systems and hence their implementation is in the interest of society at large. Therefore, the well established concept of social cost–benefit analysis (CBA) is the overall tool box that can be applied. CBA must give an answer to a number of key questions, such as: (i) the identification and valuation of costs and benefits, i.e. what are the positive and negative impacts of the project; (ii) the choice of a valid counterfactual scenario, i.e. what are the effects of the project measured against; (iii) the definition of a realistic time horizon of the project life cycle, i.e. what is the scope and the scale of adoption of

*Corresponding author; e-mail: parvathi@ifgb.uni-hannover.de

the innovation; and finally (iv) the choice of efficiency criteria that allow the judgement of the net social benefit of the project.

Applying CBA for an organic agriculture/Fair Trade project starts with the standard welfare economic framework that assumes market equilibrium for a given commodity under conventional production conditions – coffee or tea, for example. This can be displayed in the usual market diagram as shown in Fig. 5.1a. The introduction of new standards such as organic standards and Fair Trade principles will split the market into conventional and non-conventional. The market price for organic and Fair Trade products is above the price for the conventional products, as depicted in Fig. 5.1b, because marginal costs of supply are higher due to the standard. At the same time the market for organic and Fair Trade products takes away demand from the conventional market, as some consumers prefer to purchase the goods with a higher environmental and social standard. This results in a backward shift of demand in the market for conventional commodities (from D_1 to D_2) in Fig. 5.1a. However, the total market volume is likely to expand over the status quo because of consumers' willingness to increase their per capita consumption given the possibility to purchase environmentally and socially more benign products. Therefore, overall welfare will increase when summing up producer and consumer rents in both markets.

The second effect of the introduction of standards is that these are produced with lower external costs as conventional products. This is illustrated in the lower panels of Fig. 5.1a and Fig. 5.1b, respectively. Hence there is a net saving in environmental and other social costs due to the reduction of conventional produce. In addition to these short-term effects there can be more societal gains in the long run. For example, in Fair Trade systems the price premium is used to improve infrastructure and social conditions of labourers and small-scale farmers in poor communities. This can generate additional growth and poverty reduction effects (e.g. Parvathi and Waibel, 2016). Furthermore, the existence of alternative production techniques and Fair Trade standards may stimulate conventional producers to gradually change and adopt cleaner practices. Conventional food-processing companies may also change their marketing and management practices by contributing to development activities in poor communities as part of their corporate social responsibility concept. These effects are represented in an upward turn of the marginal external costs of conventional production (and trade) displayed in the lower panel of Fig. 5.1a.

The theoretical framework outlined above can serve as guidance to address the gaps in the economics of organic and Fair Trade agriculture literature. As illustrated by Fig. 5.1, a comprehensive assessment of these systems may start out with the conventional welfare economics framework. While it may not be possible in all cases to work on the basis of empirically verified supply and demand curves and to estimate consumer and producer surplus, one can establish good indicators to measure the costs and benefits of organic and Fair Trade production. These should include economic indicators and environmental as well as social indicators, which will be discussed in the following section.

5.2.1 Economic indicators

Assessing the economic performance of organic and Fair Trade systems can be done following the methodology that is generally applied when assessing the impact of innovations or policy change. Normally one would start with the assessment of productivity, financial returns and costs at the enterprise level. Since there are no specific requirements for organic and Fair Trade in applying these methodological tools as compared with conventional systems in this chapter, no further discussion will be devoted to these procedures. Instead the focus will be on discussing the methodology to measure the welfare effects particularly of small-scale producers in developing countries. For the case of organic agriculture and Fair Trade systems, additional effects that are expected to add to their benefits must be taken into account. Firstly, for Fair Trade: a major goal is reducing poverty and increasing equity. Hence static and dynamic poverty measures must be included in the analysis. Secondly, for organic agriculture: human health, natural resources and environmental effects should be included. This requires using tools of non-market evaluation methods such as environmental accounting, hedonic price methods or contingent valuation.

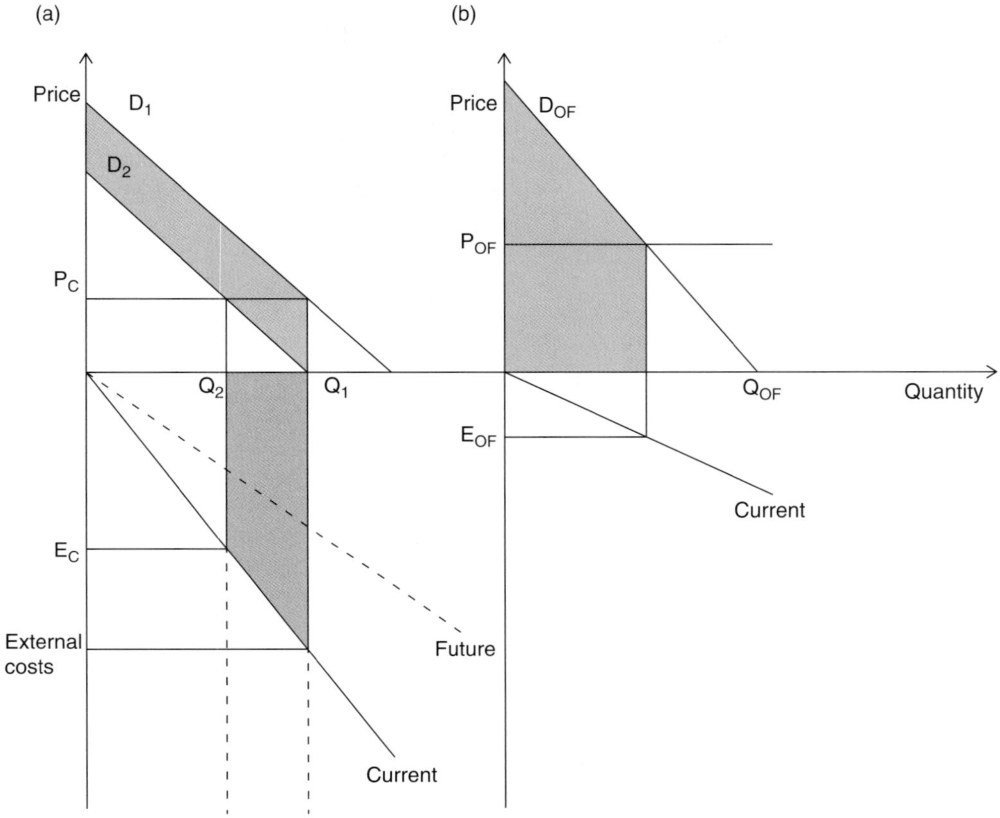

Fig. 5.1. Conceptual framework for economic studies of organic agriculture and Fair Trade systems: (a) conventional; (b) organic/Fair Trade.

For both Fair Trade and organic farming, it is necessary to take a long-term perspective and consider the sustainability dimension.

In the first step of impact assessment the analysis will start with measuring the effects at household level. There are three major economic measures that can be used as performance indicators: consumption, income and assets (e.g. Friedman, 1957; Hagenaars, 1986; Ringen, 1988; Carter and Barrett, 2006). Generally, income would be the preferred measure, though in developing countries consumption is often easier to implement. This is due to the difficulty in capturing the value of home production, informal self-employment income and seasonal labour in developing countries. In any case, the disadvantage of both measures is that they are imperfect predictors of true economic welfare, especially when shocks come into play, which is often the case in developing-country agriculture. Both measures say nothing about whether the target group can improve its capabilities to sustain the observed positive difference in income and consumption. Therefore, assets can be used to predict future income and consumption and are a more meaningful indicator of true wealth. Assets bring a dynamic perspective into the analysis. As pointed out by Carter and Barrett (2006), asset growth is a good indicator if one wants to assess long-term welfare. Hence, when evaluating the welfare of organic and Fair Trade black pepper certification on smallholder households, we use all three welfare indicators, namely total household income, consumption expenditures and assets, as dependent variables to measure current and long-term impact (Parvathi and Waibel, 2016).

5.2.2 Impact assessment methodology

Impact assessment can be done as ex-post or ex-ante evaluation. Ex-post can rely on observation of actual impact, while ex-ante must use predictions. However, predictions are also necessary in ex-post evaluation as the analysis is often done before the end of the project life cycle.

Impact assessment faces three major challenges. Firstly, we need to identify the impact pathway clearly in order to be able to attribute the effect to the intervention and this avoids attribution errors. Secondly, we must define a realistic counterfactual in order not to under- or overestimate the impact. Thirdly, we need a fair understanding of the adoption process, which requires good adoption models. Since technology adoption is a dynamic process, ideally an experimental design is needed. In the classic double delta approach, one would establish a treatment (adopter) and control (non-adopter) group, establish baseline information prior to the intervention and repeat the survey some reasonable time after the intervention. Although many impact assessment studies are carried out using cross-section data, such research lacks scientific rigour as the correlation coefficients of the adoption regressions cannot be interpreted as causal effects. If, for example, farm size is found to be a significant factor influencing adoption, this does not automatically mean that larger farmers are more likely to adopt, because current farm size could also be a consequence of earlier successful adoption decisions. Furthermore, unobserved variables (like skills) may be correlated with other explanatory variables, which can result in spurious correlations. Hence, adoption studies of Fair Trade and organic agriculture should use panel data even in the absence of complete experimental designs, in order to obtain unbiased coefficients with consistent estimates of the adoption drivers. Panel data can help to overcome the problem of endogeneity due to unobserved heterogeneity (e.g. Besley and Case, 1993; Barham *et al.*, 2004). Panel models can account for spurious causality in the adoption decisions and establish direction of causality in adoption analysis (Besley and Case, 1993). As pointed out by Doss (2006), adoption processes can be better examined by following adoption decisions over time. Furthermore, panel data allow for controlling heterogeneity across households and thereby accounting for endogenous regressors. Hence, the robustness of adoption models can be improved using panel data, even if no dis-adoption or late adoption is observed in the sample and the variability is only captured by the explanatory variables (Parvathi and Waibel, 2015).

The classic adoption model of Rogers (1995) assumes that adoption follows an S-shaped diffusion path in which the adoption dynamics depends on the differences across adopter categories. Hence, on the basis of this foundation, we can conclude that a panel model is more precise to identify adoption determinants as compared with adoption models that must rely on cross-section data. Adoption models using cross-section data usually apply logit or probit models (e.g. Burton *et al.*, 1999; Koesling *et al.*, 2008; Khaledi *et al.*, 2010) as a first step. Then, propensity score matching is used to identify the treatment effect of the intervention on the adopters. These procedures clearly suffer from the shortcomings described above and the likelihood of attribution errors is high. The core of a scientifically rigorous impact assessment method is to conduct a good counterfactual analysis.

A counterfactual group can be effectively created using a two-stage modelling framework. In the first stage, a multinomial logit selection equation can be estimated in the case of simultaneous assessment of technologies, as is the case for Fair Trade cum organic farming. An outcome equation is estimated with the multinomial selection bias correction terms calculated from the multinomial logit model entering the equation as generated regressors. In the second stage a counterfactual analysis is implemented and the average treatment effects on the treated (ATT) and the average treatment effects on the untreated (ATU) are calculated to ascertain the impact of certification on welfare. Following Di Falco and Veronesi (2013) and Teklewold *et al.* (2013), this modelling procedure which can correct for selection bias is commonly called a multinomial endogenous switching regression model and we recommend this procedure as the standard in impact assessment (Parvathi and Waibel, 2016).

5.2.3 Externalities

Some of the impacts of organic farming and Fair Trade are unlikely to be captured in market prices and therefore require additional analysis that goes beyond the standard market-based framework. Especially for organic farming, some of the benefits are the reduction of environmental and health externalities. There are generally three categories of evaluation methods (Fig. 5.2). The first one is the damage function approach where market prices are used to measure loss of productivity, preventive expenditures or mitigation costs. In practice, this is environmental accounting applied to identifiable environmental or human health effects that are not captured by product or factor markets due to institutional failure. An example is the revival of beneficial organisms for pest control (natural enemies) after adoption of organic farming.

The second group of measures falls into the category of revealed preferences using surrogate goods. Most prominently this refers to the hedonic pricing method where the non-market environmental good is a trait of a marketable good and regression analysis is used to estimate the marginal effect of altering the level of the trait (e.g. clean air on housing prices). Related methods are the wage differential approach suitable to value occupational health risks and the travel cost method applied to value environmental amenities such as landscapes.

The third category is stated preference approaches for non-market goods. These include contingent valuation methods and contingent choices, among others (Fig. 5.2). Most of these methods are survey-based approaches that establish a simulated hypothetical market where participants reveal their preferences.

The most common among these methods is contingent valuation (CV). The underlying

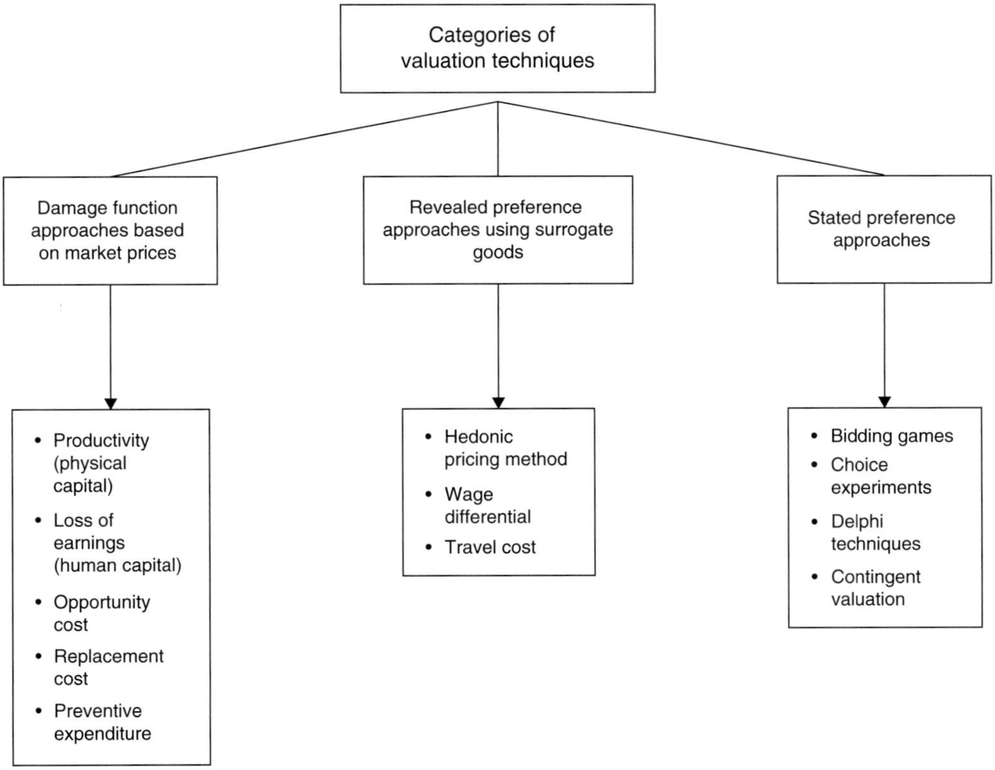

Fig. 5.2. Overview of evaluation techniques for environment and natural resources (Hufschmidt *et al.*, 1983; Bann, 2002; modified).

theoretical basis for contingent valuation is welfare economics, with utility theory at its core. In CV, a constant individual utility is taken as the basis to evaluate a change in the supply of a non-market good applying the concept of Hicks compensated demand functions. The approach is demonstrated in Fig. 5.3 for the case of evaluating pesticide-induced human health effects. The utility of the farm household (U_0) is represented as the sum of health (H_0) and other goods, summarized as income (I_0). If supply with health is improved to H_1, as would be the case if organic farming were adopted, keeping income constant ($I_0 = I_1$), an individual will move to a higher utility level (U_1). The value of the improvement in health is measured as the amount of income that the farmer is willing to forgo (willingness to pay (WTP)) in order to be as well off as before the change in health, i.e. to remain on his initial utility level U_0 with H_1, I_2, the compensating variation (C).

The elicitation of WTP is based on surveys, where respondents evaluate the non-market good in hypothetical market situations. Since CV relies on stated preferences instead of behaviour observed in real markets, it is necessary to perform a number of validity tests before accepting the results. These include, for example, the following (Mitchell and Carsson, 1989):

- Content validity, which is achieved through careful design of the survey instrument. The definition of the good and the scenarios should ensure that all important variables are included and that the correct values are measured. Bias can result if the amount of benefits of the scenario is not clear or if the way of payment is not plausible to respondents. Careful survey design pre-tests and focus group discussions are tools to enhance content validity before starting the main data collection.
- Convergent validity compares valuations of the same good obtained by different measures. If the values are correlated and tend to converge, they are assumed to be valid. However, in a specific application, it may be difficult to obtain other measures, as CV is usually applied in cases where, say, market-based prices are not available.
- Theoretical validity applies the concept that the demand for non-market goods follows the same rules as the demand for market goods.

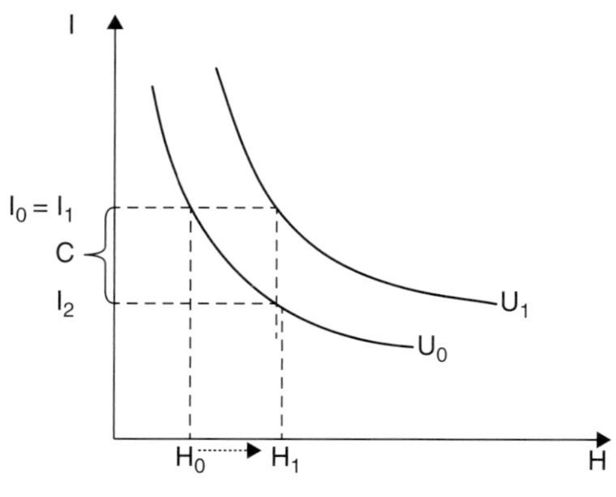

Fig. 5.3. Compensating variation for an improvement in pesticide-related health (Garming and Waibel, 2009).

If carefully done, CV (and contingent choice) can be a powerful tool for the evaluation of non-market effects that can play an important role in organic farming assessment.

5.2.4 Poverty analysis

One of the major reasons for the introduction of Fair Trade in agricultural commodities is to reduce poverty. Therefore, any analysis of these systems must assess the impact on poverty. The respective measurement tools are the Foster Greer Thorbecke (FGT) measures (see Eqn 5.1). The FGT measures allow assessment of poverty at different levels of sensitivity which is expressed in the parameter α. For α = 0 we get the headcount ratio, α = 1 is the poverty gap and higher numbers of α increasingly capture poverty severity taking into account the distribution of wealth among the population below the poverty line. Fair Trade is supposed to do both, namely lifting poor farmers and farm labourers above the poverty line as well as reducing severe poverty and the poverty gap by making wealth and access to growth opportunities more equitable.

$$P_\alpha(z) = \frac{1}{N}\sum_{i=1}^{N}\left[\frac{z - y_i}{z}\right]^\alpha \qquad (5.1)$$

where N = size of the population, z = poverty line and y_i = income of those below the poverty line.

While the FGT measures will give some information about the poverty reduction effect, a disadvantage is that they are ex-post measures that can say very little about long-term and future wealth of the target population.

Often the poor are confronted with external shocks that can take them back into poverty. Hence, the analysis should include dynamic poverty measures, namely vulnerability to poverty. A good forward-looking measure of vulnerability to poverty is the Calvo and Dercon (2005) measure. It is defined as expected deprivation accounting for probabilities of the occurrence of future negative events (e.g. see Klasen and Waibel, 2013) and is expressed in Eqn 5.2:

$$V^* = V(z, y, p) \tag{5.2}$$

where z = benchmark (i.e. the poverty line), y = vector of outcomes across n states of the world and p = a vector of corresponding probabilities.

To implement this forward-looking measure of the impact of Fair Trade or organic farming on poverty reduction, it is necessary to collect data on the likelihood of future risks and their impact on household welfare. Hence, the likelihood that a household will fall under the poverty line can be established as an outcome indicator of Fair Trade and organic projects.

5.2.5 Gender differentiation

To complete the assessment of Fair Trade and organic agriculture initiative, it is imperative that the effects on gender are measured. This means two things: (i) the share of the benefits that accrue to women; and (ii) the benefits that women-headed households achieve relative to those of male-headed households. For the first case, a gender differentiation of the benefit indicators has to be added to the study. In the second case, a comparison between female-headed and male-headed households can be undertaken and causality can be established whether female-headed households are better or worse off than their male counterparts.

An example is the study of Klasen *et al.* (2015) which investigates the welfare conditions of female-headed households. The study uses a socio-economic panel dataset with 4000 households in Thailand and Vietnam for 2007 and 2008. The study uses a pooled dataset of both waves and applies a consumption model estimated separately for each country to answer the question of whether female-headed households have lower levels of consumption compared with their male counterparts. It includes information on household characteristics and endowments. By means of a Blinder–Oaxaca decomposition, coefficient effects can be separated from endowment effects for both female- and male-headed households. The paper also estimated the Calvo–Dercon measure of household-specific vulnerability to poverty for both female- and male-headed households. Similar models can be developed for Fair Trade agriculture to provide a gender-specific impact analysis.

5.3 Summary

The purpose of this selective overview and brief description of tools for the welfare assessment of organic and Fair Trade agriculture was to equip analysts with a framework to approach such studies. The main points of this chapter are that a conceptual framework is applied that captures not just the classic market-based welfare effects but at the same time also takes into account the non-market effects in terms of environment and equity. For outcome indicators, good data must be collected on consumption, income and assets, preferably at household level. However, data should allow gender differentiation in order to be able to capture equity effects.

Impact assessment methodology must be rigorous, including a good adoption model preferably using panel data and a clearly defined impact pathway including a well defined counterfactual in order to minimize the attribution error. The modelling procedure must account for selection bias and applying a multinomial endogenous switching regression model is seen as the best choice that allows the ATT and ATU to be measured for a better understanding of the impacts.

Assessment of externalities is mandatory for a complete valuation of the benefits of organic farming and the various methods of environmental economics can be applied. Most prominent would be methods based on hypothetical markets. These are deemed most suitable when new products are entering a market and market volumes are small. Well designed survey instruments and rigorous testing procedures must be applied to assure validity of results.

Finally, poverty assessment is a needed component in the assessment of Fair Trade regimes. It is not sufficient to remain with the static FGT poverty measures such as head count ratio and poverty depth but criteria of dynamic poverty such as vulnerability to poverty should be added as a forward-looking measure of the reduction of expected deprivation. Last but not least, a state-of-the-art analysis of either organic farming or Fair Trade must include gender dimension as one important aspect of equity.

References

Bann, C. (2002) An overview of valuation techniques: advantages and limitations. *Asean Biodiversity* 2(2), 8–16.

Barham, B.L., Foltz, J.D., Jackson-Smith, D. and Moon, S. (2004) The dynamics of agricultural biotechnology adoption: lessons from series rBST use in Wisconsin, 1994–2001. *American Journal of Agricultural Economics* 86(1), 61–72.

Besley, T. and Case, A. (1993) Modeling technology adoption in developing countries. *The American Economic Review* 83(2), 396–402.

Burton, M., Rigby, D. and Young, T. (1999) Analysis of the determinants of adoption of organic horticultural techniques in the UK. *Journal of Agricultural Economics* 50(1), 47–63.

Calvo, C. and Dercon, S. (2005) *Measuring Individual Vulnerability*. (Discussion Paper Series No. 229), Department of Economics, Oxford University, Oxford, UK.

Carter, M.R. and Barrett, C.B. (2006) The economics of poverty traps and persistent poverty: an asset-based approach. *Journal of Development Studies* 42(2), 178–199.

Di Falco, S. and Veronesi, M. (2013) How can African agriculture adapt to climate change? A counterfactual analysis from Ethiopia. *Land Economics* 89(4), 743–766.

Doss, C.R. (2006) Analyzing technology adoption using microstudies: limitations, challenges, and opportunities for improvement. *Agricultural Economics* 34(3), 207–219.

Friedman, M. (1957) *A Theory of the Consumption Function*. Princeton University Press, Princeton, New Jersey.

Garming, H. and Waibel, H. (2009) Willingness to pay for health risks from pesticides in developing countries: a case study from Nicaragua. *European Journal of Health Economics* 88(2), 499–511.

Hagenaars, A.J. (1986) *The Perception of Poverty*. North-Holland, Amsterdam.

Hufschmidt, M.M., James, D.E., Meister, A.D., Bower, B.T. and Dixon, J.A. (1983) *Environment, Natural Systems, and Development: an Economic Valuation Guide*. John Hopkins University Press, Baltimore, Maryland.

Khaledi, M., Weseen, S., Sawyer, E., Ferguson, S. and Gray, R. (2010) Factors influencing partial and complete adoption of organic farming practices in Saskatchewan, Canada. *Canadian Journal of Agricultural Economics/Revue canadienne d'agroeconomie* 58(1), 37–56.

Klasen, S. and Waibel, H. (eds) (2013) *Vulnerability to Poverty: Theory, Measurement and Determinants, with Case Studies from Thailand and Vietnam*. Palgrave Macmillan, New York.

Klasen, S., Lechtenfeld, T. and Povel, F. (2015) A feminization of vulnerability? Female headship, poverty, and vulnerability in Thailand and Vietnam. *World Development* 71 (July), 36–53. doi: 10.1016/j.worlddev.2013.11.003.

Koesling, M., Flaten, O. and Lien, G. (2008) Factors influencing the conversion to organic farming in Norway. *International Journal of Agricultural Resources, Governance and Ecology* 7(1–2), 78–95.

Mitchell, R.C. and Carson, R.T. (1989) *Using Surveys to Value Public Goods: the Contingent Valuation Method*. Resources for the Future, RFF Press, Washington, DC.

Parvathi, P. and Waibel, H. (2015) Adoption and impact of black pepper certification in India. *Quarterly Journal of International Agriculture* 54(2), 133–161.

Parvathi, P. and Waibel, H. (2016) Organic agriculture and fair trade: a happy marriage? A case study of certified smallholder black pepper farmers in India. *World Development* 77, 206–220.

Ringen, S. (1988) Direct and indirect measures of poverty. *Journal of Social Policy* 17(03), 351–365.

Rogers, E.M. (1995) *Diffusion of Innovations*, 4th edn. The Free Press (Simon and Schuster), New York.

Teklewold, H., Kassie, M., Shiferaw, B. and Köhlin, G. (2013) Cropping system diversification, conservation tillage and modern seed adoption in Ethiopia: impacts on household income, agrochemical use and demand for labor. *Ecological Economics* 93, 85–93.

6 Consumer Behaviour in the Organic and Fair Trade Food Market in Europe

Katrin Zander[1]*, Rosa Schleenbecker[2] and Ulrich Hamm[2]
[1]*Thünen Institute of Market Analysis, Federal Research Institute for Rural Areas, Forestry and Fisheries, Braunschweig, Germany;* [2]*Department of Agricultural and Food Marketing, University of Kassel, Germany*

6.1 Introduction

The production of organic and Fair Trade products is usually aligned with higher production costs, which have to be covered somewhere in the supply chain. From earlier research it is well known that in Western markets consumer segments exist that are willing to ask for organic and Fair Trade products even if they are more expensive (Krystallis *et al.*, 2006; Urena *et al.*, 2008; Corsi and Novelli, 2011; Liljenstolpe, 2011; Zander *et al.*, 2013; Rödiger and Hamm, 2015). Organic production and Fair Trade are process qualities that are not verifiable by consumers, either before or after purchase and consumption. In information economics this type of good is called 'credence good' (Caswell and Anders, 2011; Janssen and Hamm, 2012). It is characterized by so-called 'information asymmetry', since food suppliers know the production processes while consumers do not.

Preconditions for consumers' higher willingness to pay are that they know about the specific product qualities, that they appreciate them and that they trust the information provided about them. That is why consumers' knowledge and trust need to be at the centre of the discussion about prospects of the market for organic and Fair Trade products. Any effort to increase market shares for these product groups should start by asking what consumers want with respect to the specific product qualities. With this knowledge, food suppliers are able to actively address consumers and activate their willingness to pay.

The aim of this chapter is to give an overview of consumer preferences for organic and Fair Trade food and to demonstrate the impacts on standard setting, certification, labelling and communication. It is about: (i) opening the view to the relevance of having the consumers' perspective in mind when discussing perspectives and opportunities of Fair Trade and organic markets; and (ii) better understanding consumer behaviour in the ethical food market.

Questions to be addressed in this chapter are:

- Why do European consumers buy organic and Fair Trade products?
- What do we know about consumers' willingness to pay for organic and Fair Trade products?
- What do we know about consumers who buy organic and Fair Trade products?
- Why do European consumers not buy organic and Fair Trade products?

*Corresponding author; e-mail: katrin.zander@thuenen.de

- How should standard setting, certification, labelling and communication be designed to create consumers' trust in the process-related credence goods?

The chapter is structured as follows. Firstly, results of European consumer research on their preferences, buying motives, willingness to pay as well as barriers for not buying organic and Fair Trade products are summarized. Secondly, the implications for standard setting, certification, labelling and communication are outlined. The chapter closes with some conclusions for the organic and Fair Trade supply chain actors.

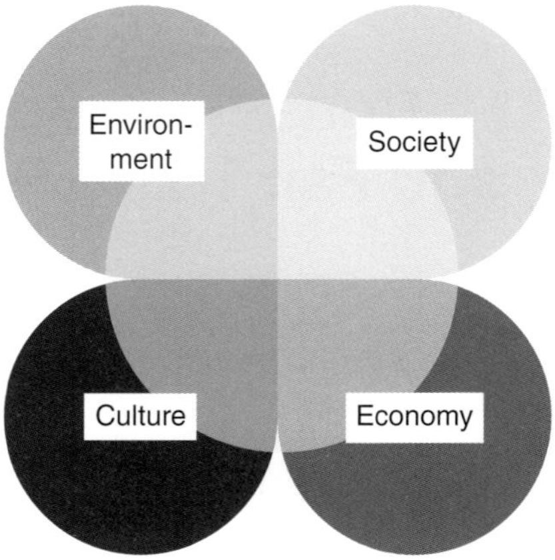

Fig. 6.1. Pillars of sustainability in the food sector.

6.2 Consumer Preferences for Organic and for Fair Trade Food

The demand for organic and Fair Trade products is rooted in the phenomenon that consumers in countries with a high economic wealth are considering not only material aspects of the food when shopping, such as nutritional value, taste and freshness, but also aspects that are associated with production processes. Local production as well as organic and Fair Trade products are typical examples. This phenomenon is called ethical consumer behaviour. Ethical consumer behaviour implies that consumers orient their buying behaviour at ethical product properties.

Ethical concerns are traditionally categorized according to the three pillars of sustainability that comprise environmental, social and economic issues. However, none of these categories seems to be adequate to include farm animal welfare and traditional production methods and landscape preservation. Thus, an additional fourth pillar might to be needed for all these issues (Michelsen, 2014) (Fig. 6.1).

In the context of food production the pillars contain the following aspects:

- environmental, e.g. responsible use of natural resources, including water, soil and biodiversity;
- social, e.g. fair, safe and equitable working conditions, banning of child labour, employment of disabled people;
- economic, e.g. fair prices for farmers, manufacturers or retailers, long-term contracts for less powerful members of supply chains, fair trading conditions, support for enterprises in disadvantaged or mountainous regions; and
- cultural and other, e.g. farm animal welfare, preservation of cultural heritage such as specific (e.g. artisan) production and manufacturing techniques or specific landscape features and local production.

The boundaries of these categories are not stable and some of the aspects are cross-cutting. 'Local production' may serve as an example, since environmental issues, such as short transport distances, as well as economic and cultural issues are associated with the term.

The market shares for organic and Fair Trade food are steadily increasing, with substantial growth rates during recent years. The Fair Trade food market in Germany increased by almost a factor of 20 between 2003 and 2015 and was at about €1 billion in 2015 (Statista, 2016a). During the same period, the organic market in Germany has grown by 180% from €3.1 billion to €8.6 billion (Statista, 2016b). However, market shares are still small for both segments: in 2014 the share of organic food in all retail sales was at about 4.4% in Germany and ranged in Europe from 0.2% in Portugal and Poland to 7.6% in Denmark in 2014 (Schaack and Rampold, 2016). The Fair Trade food segment is even smaller and was at about €1 billion in Germany in 2015 (Statista, 2016b). In Switzerland the Fair Trade market is at about 600,000 Swiss francs (Fr) and the organic market accounts for Fr2.3 billion (Bio Suisse, 2016; Max Havelaar Foundation,

2016). In the UK the Fair Trade market has a volume of about £1.6 billion and the organic market is at £1.95 billion (Fair Trade Association, 2016; Soil Association, 2016). All numbers presented refer to the food and beverages market, which constitutes the major part of Fair Trade products. Non-food products are mainly craftworks and textiles. Recently, the sales growth of textiles is tremendous. Nevertheless, the share of food and beverages is still high; in Germany, for example, it is about 75% (Forum Fairer Handel, 2014).

Even though market shares are small, organic as well as Fair Trade products have entered the mainstream market. Nowadays, organic and Fair Trade products can be found in supermarkets and discount stores next to their traditional sales channel, the worldshops and wholefood/organic stores. Additionally, the online sales channel is continuously growing and online grocery shopping has started to gain importance (Morgan Stanley, 2016). Consequently, Fair Trade and organic products can also be bought online, whether groceries or other products such as fashion (cf. Fair Trade, 2016). In Germany, online sales volumes of Fair Trade products doubled in 2013 (Forum Fairer Handel, 2014).

Due to the fact that they address similar consumer groups, the organic and the Fair Trade market segments are interrelated, which is reflected in a growing share of products that are certified according to both Fair Trade and organic standards. A specific aspect that distinguishes Fair Trade products from organic products is the fact that 'fair trade' it is not legally defined. Consequently, a variety of labels and brands exist, among which the private Fair Trade Labelling Organizations (FLO) standard is the most prominent one in the European market. Examples of other labels indicating that their products are produced under 'fair' conditions are 'fair for life' or UTZ or the Gepa-logo of the importer Gepa.

6.2.1 Buying motives for organic and Fair Trade food

Organic and Fair Trade food comprises various characteristics that differ from conventional production.

In the case of organic products, these are environmental protection and less residues (due to the non-use of chemical pesticides and easily soluble fertilizers) or higher animal welfare standards, among others. Organic consumers have been observed to be increasingly dissatisfied with anonymous, homogeneous food, which may be produced under unknown social conditions. They want greater traceability and information about the diverse origins and conditions under which food is produced, and from where and how it is transported.

The characteristics of Fair Trade products are the production under more 'fair' working and trading conditions, such as the payment of minimum prices which are supposed to cover the costs of production, pre-financing of the harvest and the financial support of social projects that target the producers and their families.

From the marketing point of view it is important to know which of the different aspects consumers are interested in. In the case of organic products, expectations are related to the environment, animal welfare, support for local production structures and the well-being of those engaged in food production (Browne *et al.*, 2000; Torjusen *et al.*, 2004; Ozcaglar-Toulouse *et al.*, 2006; Aschemann and Hamm, 2007; Goig, 2007; Hughner *et al.*, 2007; Lusk and Briggeman, 2009).

Motives for buying organic food can be grouped into hedonic/egoistic and altruistic reasons (Table 6.1). For most of the consumers a bundle of different reasons is given as they value egoistic as well as altruistic product properties (Padel and Foster, 2005; Aertsens *et al.*, 2009; Hjelmar, 2011; Padilla Bravo *et al.*, 2013).

Table 6.1. Egoistic and altruistic reasons for buying organic food.

Egoistic motives	Altruistic motives
Health	Environmental protection
Taste	Animal welfare
Less residues (pesticides, fertilizers, additives, GMO-free)	Naturalness
Nourishing	GMO-free
Food safety	(Protection of small farms and rural communities)
	(Local production)

Research results are contradictory with respect to predominant reasons for buying organic food. While some studies stress egoistic reasons (e.g. Wier and Calverley, 2002; Zanoli and Naspetti, 2002; Aertsens *et al.*, 2009; Zagata and Lostak, 2012; Meyer-Höfer *et al.*, 2014), others found that altruistic motives are more important (Aarset *et al.*, 2004; Schleenbecker and Hamm, 2013).

Chapter 8 contains a case study giving insights into consumer preferences for Fair Trade products and also highlights the purchase motives for Fair Trade products. The study shows that Fair Trade consumers purchased Fair Trade products firstly because they wanted to make a contribution to more equity in the world. Secondly, participants stated that they wanted to support the producers. Thirdly, they stated that they purchased Fair Trade for the good feeling. As in the case of purchasing organic products, consumers' motives for purchasing Fair Trade products are both hedonistic and altruistic.

6.2.2 Willingness to pay

Competition in food markets is to a large degree a question of prices. That is why consumer willingness to pay (WTP) is important when aiming at market differentiation. Research has shown that consumers are willing to pay significant price premiums for food that is labelled as organic or Fair Trade. The amount of the additional WTP varies widely and depends on the product group (e.g. Hemmerling *et al.*, 2015). It is higher for fruit and vegetables, meat, cheese and bread, while it has been found to be lower for milk, cereals and potatoes (Mondelaers *et al.*, 2009; Van Loo *et al.*, 2013; Rödiger and Hamm, 2015). Studies report additional WTP of consumers of about 22% for organic milk (Aldanondo-Ochoa and Almansa-Sáez, 2009) and organic yoghurt (Van Loo *et al.*, 2013), 50% for organic carrots (Mondelaers *et al.*, 2009), about 60% for organic apples and 120% for organic eggs (Janssen and Hamm, 2014), 16% (Brugarolas *et al.*, 2005) and 65% (Krystallis *et al.*, 2006) for organic wine and 10% for organic coffee (Langen, 2011).

A number of studies have surveyed the willingness to pay for Fair Trade products. However, the results differ. Whereas McCluskey *et al.* (2009) found an additional WTP of more than 70% for Fair Trade bananas in the USA, Pelsmacker *et al.* (2005) could only detect an additional WTP for Belgian consumers of 10% for Fair Trade products in general. Loureiro and Lotade (2005) found an additional WTP of 3.3% for Fair Trade coffee in the USA. WTP also largely differs between consumer segments, which is why it is not helpful to argue with an average WTP for organic or Fair Trade products; some consumers are not interested in buying organic food, resulting in a WTP of zero, driving the average WTP downwards. Higher WTP is usually correlated with a favourable attitude towards organic food and with higher frequencies of buying organic food, which implies that some consumers do not care too much about prices if they really are convinced of the higher quality.

6.2.3 Who is buying organic and Fair Trade food?

The question of the characteristics of the typical organic consumer is relevant when thinking about addressing them in an effective manner by product design and attributes. Socio-demographic characteristics were found to have some impact: organic food consumers are described as better educated, middle-aged, wealthier, more frequently female, mostly having children (e.g. Hughner *et al.*, 2007; Aertsens *et al.*, 2009; Hjelmar, 2011) and better informed (Vermeir and Verbeke, 2006). However, socio-demographic characteristics are not sufficient for segmentation, and relevant attitudes, behavioural characteristics and beliefs are further decisive factors. Generally, organic consumers are characterized by concern about animal welfare, the natural environment and local production as well as by a strong interest in pro-social behaviour. Consumers with a higher commitment are convinced of the higher quality and better taste of organic food (Hjelmar, 2011).

Different consumer segments exist and these differ in their responsiveness to organic and Fair Trade products. With respect to organic consumers, mostly two to three different groups are identified according to the frequency or intensity of organic purchases. 'Occasional' and 'regular' or 'frequent' buyers are distinguished (Janssen and Hamm, 2014; Zander *et al.*, 2015) and sometimes an intermediary group is defined as well (e.g. Van Loo *et al.*, 2013; Janssen and Hamm, 2014). Hjelmar (2011) distinguishes between politically/ethically minded consumers and pragmatic consumers, Mondelaers *et al.* (2009) between heavy and light users and Hempel and Hamm (2016) between organic-minded and non-organic-minded consumers. There might be some overlapping between heavy, regular and politically/ethically minded consumers on the one hand and light, occasional and pragmatic consumers on the other hand. From the marketing perspective it is important that pragmatic consumers switch easily to conventional products when their expectations about product quality are not fulfilled.

6.3 Why Consumers Do Not Buy Organic and Fair Trade Products

Although the concepts of sustainability and ethical conviction have experienced broad social acceptance, the market shares of ethical products are small (see above). Thus, being ethically minded does not necessarily imply ethical buying behaviour.

The fact that attitudes are not directly transferred into action at the point of sale has been addressed as 'attitude–behaviour gap' (e.g. Carrington *et al.*, 2010; Solomon *et al.*, 2010). This phenomenon can also be observed with regard to organic food purchases (Padel and Foster, 2005; Vermeir and Verbeke, 2006; Feldmann and Hamm, 2015; Hemmerling *et al.*, 2015) as well as to Fair Trade product purchases (Pelsmacker *et al.*, 2006; Vantomme *et al.*, 2006) or ethical consumption in general (Carrington *et al.*, 2010; Papaoikononou *et al.*, 2011).

Some of the several reasons why behaviour may differ from general attitudes are outlined below.

6.3.1 Low knowledge

Low knowledge of 'ethical' production standards is a barrier for ethical buying behaviour, since knowledge is one of the preconditions for translating attitudes into behaviour. Without knowing about production processes and product properties, consumers are not able to adjust their buying behaviour in favour of ethical products and according to their individual ethical preferences (Carrington *et al.*, 2010). Earlier research showed that consumers' knowledge on standards and certification, even in well established market segments such as organic food and Fair Trade, is often limited (e.g. Hughner *et al.*, 2007; Padel, 2010; Janssen and Hamm, 2012; Zander and Zanoli, 2013; Hemmerling *et al.*, 2015). In many cases, there is some basic understanding, for example with respect to the use of pesticides. Nevertheless, consumers have their own expectations about organic farming practices and follow their intuitive quality expectations instead of relying on factual information about production standards. These 'intuitive' quality expectations are not necessarily in line with real-world production, yet consumers expect them to be met.

6.3.2 Information

Information is the most common way to tackle lack of knowledge. Consumers are faced with huge amounts of information in day-to-day life. While formerly the availability of information and the cost of information acquisition were limiting factors in information search, the task for consumers today is to actively select the information relevant to their individual decision making and to keep away from irrelevant information. They need to control their individual information flow in order to arrive at reasonable decisions, by deciding not to look for information they perceive not to be relevant. In this respect they use mental short-cuts, particularly in habitual decision making such as food choices.

In a world full of information, 'information overload' is a frequently observed phenomenon in contemporary decision making and is characterized by a situation in which consumers feel overburdened or even confused because of the huge amount of available information (Hwang and Lin, 1999; Verbeke, 2005). As a consequence, consumers are no longer able to make buying decisions according to their individual preferences. This may lead to wrong decision making in the sense of not choosing a product that is in line with their own preferences or to give up due to the feeling of not being able to make the 'right' decision (Hjelmar, 2011).

6.3.3 Credibility and trust

Credibility and trust are particularly important in ethical market segments because of the pronounced 'information asymmetry'. Since consumers cannot verify process qualities such as organic and Fair Trade, credibility and trust are crucial for market success. Lacking trust in standard setting, certification and labelling has frequently been reported by consumers when asked for their reasons for not buying 'ethically'. This holds true even for organic products where production standards are well defined and certification systems work well (e.g. Padel and Foster, 2005; Janssen and Hamm, 2012).

6.3.4 High prices

High prices are frequently mentioned as a reason for not buying organic food (e.g. Padel and Foster, 2005). Usually, ethical products are more expensive than conventional products and consumers have to trade off ethical product properties against their individual budgets. Given the low price knowledge of many consumers (Rödiger *et al.*, 2016), this reason for not buying organic food is somewhat surprising. Possible explanations are that organic food has a high price image and low interest in increasing knowledge about prices keeps people away from buying organic food. Thus, perceived price differences may constitute an important barrier, particularly for new users, against buying organic products (Mondelaers *et al.*, 2009). In this respect, trust and also the perceived risk of being cheated and of not getting what is expected act as an amplifier. Even slight distrust might shift the decision towards the conventional alternative when money is short.

6.3.5 Limited availability

Limited availability of organic and Fair Trade food frequently limits its consumption. Although organic and Fair Trade products can be found in most of the food stores in Europe, the variety of products offered in the mainstream market is very low compared with conventional products and so consumers are forced to visit various shops in order to fill their shopping basket with a large array of ethical products. This contrasts with the objective of many consumers of efficient shopping that only takes a little time. Online grocery shopping has started to enter the market and includes organic and Fair Trade products, offering consumers new and convenient shopping possibilities.

6.4 Implications for Standard Setting, Certification and Communication

While certification and labelling are frequently used as synonyms, it is important to differentiate between both terms with regard to consumer communication. Certification is the process of verifying production standards. In contrast, labelling is about communicating specific production standards (and their certification) to the target group, in this case consumers. There are several lessons to learn for standard setting, certification and communication/labelling that affect consumer purchase behaviour.

6.4.1 Standard setting

When aiming to convince consumers to buy ethical products, their interests and expectations should be considered in setting production standards as well (Carrigan and Attalla, 2001). Only then will their willingness to choose ethical products and to pay price premiums be activated. Ethical production standards that aim to provoke a higher willingness to pay need to offer consumers a real 'plus', an additional benefit, compared with common production, which implies that the standards need to clearly exceed legislation. The differences should be noticeable to consumers and be relevant to their purchase decision. The specific challenge exists in that consumer knowledge is often limited (see above); nevertheless they do have some (latent) expectations that need to be considered in setting standards.

Consumer preferences supposedly change over time, a fact that affects their expectations on ethical production and should be reflected in ethical production standards. These challenges exist in the organic and in the Fair Trade markets, and the increasing share of Fair Trade products that are additionally certified for organic standards might serve as a proof of changing consumer preferences over time. However, in this case it is not that the standards themselves are adapted; instead, additional standards and certification are applied.

6.4.2 Certification

The aim of certification is to guarantee the compliance of products and production processes with defined standards and to address information asymmetry between food suppliers and consumers. Certification is needed to assure consumers that products really do meet the ethical standards promised by food suppliers. Certification processes have to be reliable and transparent and should be independent from the certified companies (Jahn *et al.*, 2005; Albersmeier *et al.*, 2010; Caswell and Anders, 2011) since independence is a precondition for credibility and is particularly relevant with credence goods (Caswell and Anders, 2011; Janssen and Hamm, 2012). With respect to organic food, standards are clearly defined by law and are mandatory for all organic food offered in the European Union (EU Regulation 834/2007). For Fair Trade products the TransFair label is the most prominent label. In order to carry the label, products need to be certified by the independent, private FLO-CERT certification agency.

6.4.3 Communication and labelling

Product properties that are aligned with production and trading processes such as organic and Fair Trade cannot usually be observed by consumers. That is why they need to be explicitly communicated to consumers in order to make them aware and to motivate them to buy. Effective communication with consumers is a big challenge in a world full of information. As described above, consumers are facing huge amounts of information in day-to-day life and so their interest in and capacity to achieve and to process additional information are scarce.

The consequence for communication of ethical product properties such as organic or Fair Trade is that only relevant and short messages should be used with the products. More and in-depth information should be made available (e.g. by the internet) in a manner easily understandable by consumers and easily accessible to consumers with better knowledge and higher interest.

Compliance with certain standards and certification (credence attributes) is usually communicated to consumers by using labels and logos (sometimes also by brands). Logos and labels have the advantage that they can be quite simple and may have a high recognition value. They are unique and consumers may be willing to repurchase products with this label if they have had good previous experiences with products carrying the label.

Credibility is crucial for every labelling scheme. Since the specific characteristics cannot be observed with the product, consumers have to trust the indications made by food suppliers. When aiming at credibility, the reference point needs to be consumers' interpretation of the terms used in communication. Care should be taken with respect to common understanding of specific terms such as 'fair' and 'sustainable' ('organic' is protected by law in the EU). Even if the terms are used correctly, consumers might understand something different and may feel cheated when they get to know the actual production standards. Consumers' expectations need to be the benchmark for every communication, since perceived or real disparity between expectations and reality may result in the perception of fraud. Shortcomings in this respect supposedly are an important reason for the frequently observed limited trust of consumers in existing labelling. The provision of reliable, honest and comprehensible information is a crucial measure to increase credibility.

As described above, consumer knowledge of organic production and Fair Trade standards is frequently rather low. Three consequences for communication and marketing of ethical products are worth mentioning: (i) product differentiation within the ethical market is difficult (ii) communication is very challenging; and (iii) food suppliers have to consider consumers' expectations and knowledge in order to prevent misunderstandings. Misunderstandings may easily be perceived as intents of fraud and will result in loss of credibility.

6.5 Conclusions

One important aim of standard setting, certification and communication, including labelling, in the field of ethical production such as organic and Fair Trade is access to Western markets. This chapter argues that the consumer perspective needs to be considered to a much larger extent than is currently the case. Only then will standard setting and certification open new markets by addressing new consumer segments.

Communication is crucial for credibility and also for consumer acceptance of organic and Fair Trade products as well as for their willingness to pay. It has been shown that price premiums and consumers' willingness to pay for organic and Fair Trade food differs widely (e.g. Rödiger and Hamm, 2015). In the end, high-quality products in combination with tailor-made communication can realize high price premiums that will be sufficient to cover additional production costs.

References

Aarset, B., Beckmann, S., Bigne, E., Beveridge, M., Bjørndal, T., Bunting, J., McDonagh, P., Mariojouls, C., Muir, J., Prothero, A., Reisch, L., Smith, A., Treteras, R. and Young, J. (2004) The European consumers' understanding and perceptions of the 'organic' food regime: the case of aquaculture. *British Food Journal* 106, 93–105.

Aertsens, J., Verbeke, W., Mondelaers, K. and van Huylenbroeck, G. (2009) Personal determinants of organic food consumption: a review. *British Food Journal* 111, 1140–1167.

Albersmeier, F., Schulze, H. and Spiller, S.A. (2010) System dynamics in food quality certifications: development of an audit integrity system. *International Journal on Food System Dynamics* 1, 69–81.

Aldanondo-Ochoa, A.M. and Almansa-Sáez, C. (2009) The private provision of public environment: consumer preferences for organic production systems. *Land Use Policy* 26, 669–682.

Aschemann, J. and Hamm, U. (2007) The organic market. In: Lockeretz, W. (ed.) *Organic Farming. An International History*. CAB International, Wallingford, UK, pp. 123–151.

Bio Suisse (2016) *Bio in Zahlen*. Available at: http://www.bio-suisse.ch/de/bioinzahlen.php (accessed 10 November 2016.

Browne, A.W., Harris, P.J.C., Hofny-Collins, A.H., Pasiecznik, N. and Wallace, R.R. (2000) Organic production and ethical trade: definition, practice and links. *Food Policy* 25(1), 69–89.

Brugarolas Mollá-Bauzá, M., Martínez-Carrasco Martínez, L., Martínez Poveda, A. and Rico Pérez, M. (2005) Determination of the surplus that consumers are willing to pay for an organic wine. *Spanish Journal of Agricultural Research* 3(1), 43–51.

Carrigan, M. and Attalla, A. (2001) The myth of the ethical consumer – do ethics matter in purchase behaviour? *Journal of Consumer Marketing* 18(7), 560–577.

Carrington, M., Neville, B. and Whitwell, G. (2010) Why ethical consumers don't walk their talk: towards a framework for understanding the gap between the ethical purchase intentions and actual buying behaviour of ethical minded consumers. *Journal of Business Ethics* 97, 139–158.

Caswell, J.A. and Anders, S.M. (2011) Private versus third party versus government labelling. In: Lusk, J.L., Roosen, J. and Shogren, J. (eds) *The Oxford Handbook of the Economics of Food Consumption and Policy*. Oxford University Press, Oxford, UK, pp. 472–498.

Corsi, A. and Novelli, S. (2011) Willingness-to-pay in terms of price: an application to organic beef during and after the 'mad cow' crisis. *Review of Agricultural and Environmental Studies* 92, 25–46.

Fair Trade (2016) *Online-Shopping List*. Available at: https://www.fairtrade-deutschland.de/einkaufen/online-shoppingliste.html (accessed 30 June 2016).

Fair Trade Association (2016) *Shoppers Stand behind Fair Trade through Tough Grocery Market*. Fair Trade Foundation, London. Available at: http://www.fairtrade.org.uk/en/media-centre/news/february-2016/shoppers-stand-behind-Fair Trade-through-tough-grocery-market. (accessed 6 June 2016).

Feldmann, C. and Hamm, U. (2015) Consumers' perceptions and preferences for local food: a review. *Food Quality and Preference* 40,152–164.

Forum Fairer Handel (2014) *Factsheet. Der Faire Handel in Deutschland*. Forum Fairer Handel, Berlin. Available at: https://www.forum-fairer-handel.de/fileadmin/user_upload/dateien/jpk/jpk_2014/factsheet_ffh_2013.pdf (accessed 30 June 2016).

Goig, R.L. (2007) Trade and global cognitive orientation – a focus on Spanish fair trade consumers. *International Journal of Consumer Studies* 31, 468–477.

Hemmerling, S., Hamm, U. and Spiller, A. (2015) Consumption behaviour regarding organic food from a marketing perspective – a literature review. *Organic Agriculture* 5(4), 277–313.

Hempel, C. and Hamm, U. (2016) How important is local food to organic-minded consumers? *Appetite* 96, 309–318.

Hjelmar, U. (2011) Consumers' purchase of organic food products. A matter of convenience and reflexive practices. *Appetite* 56, 336–344.

Hughner, R.S., McDonagh, P., Prothero, A., Shultz, C.J. and Stanton, J. (2007) Who are organic food consumers? A compilation and review of why people purchase organic food. *Journal of Consumer Behaviour* 6(5), 94–110.

Hwang, M.I. and Lin, J.W. (1999) Information dimension, information overload and decision quality. *Journal of Information Science* 25, 213–218.

Jahn, G., Schramm, M. and Spiller, A. (2005) The reliability of certification: quality labels as policy tool. *Journal of Consumer Policy* 28(1), 53–73.

Janssen, M. and Hamm, U. (2012) The mandatory EU logo for organic food: consumer perceptions. *British Food Journal* 114(3), 335–352.

Janssen, M. and Hamm, U. (2014) Governmental and private certification labels for organic food: consumer attitudes and preferences in Germany. *Food Policy* 49, 437–448.

Krystallis, A., Fotopoulos, C. and Zotos, Y. (2006) Organic consumers' profile and their willingness to pay (WTP) for selected organic food products in Greece. *Journal of International Consumer Marketing* 19(1), 81–106.

Langen, N. (2011) Are ethical consumption and charitable giving substitutes or not? Insights into consumers' coffee choice. *Food Quality and Preference* 22, 412–421.

Liljenstolpe, C. (2011) Demand for value-added pork in Sweden: a latent class model approach. *Agribusiness* 27, 129–146.

Loureiro, M. and Lotade, J. (2005) Do fair trade and eco-labels in coffee wake up the consumer conscience? *Ecological Economics* 53(1), 129–138.

Lusk, J.L. and Briggeman, B.C. (2009) Food values. *American Journal of Agricultural Economics* 91(1), 184–196.

Max Havelaar Foundation (2016) *Jahres- und Wirkungsbericht 2015*. Max Havelaar-Stiftung (Schweiz), Zürich. Available at: https://www.maxhavelaar.ch/was-ist-Fair Trade/max-havelaar/jahresbericht.html (accessed 6 June 2016).

McCluskey, J., Durham, C. and Horn, P. (2009) Consumer preferences for socially responsible production attributes across food products. *Agricultural and Resource Economics Review* 38(3), 345–356.

Meyer-Höfer, M.v., Nitzko, S. and Spiller, A. (2014) Is there an expectation gap? Consumers' expectations towards organic. *British Food Journal* 117(5), 1527–1546.

Michelsen, G. (2014) Education for sustainable development: status quo and perspectives. *International Yearbook for Research in Arts Education* 2, 121–129.

Mondelaers, K., Verbeke, W. and van Huylenbroeck, G. (2009) Importance of health and environment as quality traits in the buying decision of organic products. *British Food Journal* 111(10), 1120–1139.

Morgan Stanley (2016) *Are Groceries the Next Big Driver of Global eCommerce?* Available at: http://www.morganstanley.com/ideas/online-groceries-could-be-next-big-ecommerce-driver (accessed 30 June 2016).

Ozcaglar-Toulouse, N., Shiu, E. and Shaw, D. (2006) In search of fair trade – ethical consumer decision making in France. *International Journal of Consumer Studies* 30, 502–514.

Padel, S. (2010) The European regulatory framework and its implementation in influencing organic inspection and certification systems in the EU. CERTCOST Report Deliverable 11, Organic Research Centre, Newbury, UK. Available at: http://certcost.org/Lib/CERTCOST/Deliverable/D14_D11.pdf (accessed 8 May 2016).

Padel, S. and Foster, C. (2005) Exploring the gap between attitudes and behaviour. Understanding why consumers buy or do not buy organic food. *British Food Journal* 107(8), 606–625.

Padilla Bravo, C., Cordts, A., Schulze, B. and Spiller, A. (2013) Assessing determinants of organic food consumption using data from the German National Nutrition Survey II. *Food Quality and Preference* 28(1), 60–70.

Papaoikononou, E., Ryan, G. and Ginieis, M. (2011) Towards a more holistic approach of the attitude behaviour gap in ethical consumer behaviours: empirical evidence from Spain. *International Advances in Economic Research* 17(1), 77–88.

Pelsmacker, P. de, Driesen, L. and Rayp, G. (2005) Do consumers care about ethics? Willingness to pay for fair-trade coffee. *The Journal of Consumer Affairs* 39(2), 363–385.

Pelsmacker, P. de, Janssens, W., Sterckx, E. and Mielants, C. (2006) Fair-trade beliefs, attitudes and buying behaviour of Belgian consumers. *International Journal of Nonprofit and Voluntary Sector Marketing* 1, 125–138.
Rödiger, M. and Hamm, U. (2015) How are organic food prices affecting consumer behaviour? A review. *Food Quality and Preference* 43, 10–20.
Rödiger, M., Plaßmann, S. and Hamm, U. (2016) Organic consumers' price knowledge, willingness-to-pay and purchase decision. *British Food Journal* 118(11), 2732–2743.
Schaack, D. and Rampold, C. (2016) *AMI Marktbilanz Öko-Landbau 2016*. AMI, Bonn, Germany.
Schleenbecker, R. and Hamm, U. (2013) Consumers' perception of organic product characteristics. A review. *Appetite* 71, 420–429.
Soil Association (2016) *Organic Market Report 2016*. Available at: https://www.soilassociation.org/certification/market-research-and-data/organic-market-report-2016/ (accessed 6 June 2016).
Solomon, M.R., Bamossy, G., Askegaard, S. and Hogg, M. (2010) *Consumer Behaviour: A European Perspective*. Prentice Hall Europe, New York.
Statista (2016a) *Umsatz mit Fair Trade-Produkten in Deutschland in den Jahren 1993 bis 2015*. Available at: http://de.statista.com/statistik/daten/studie/226517/umfrage/fairtrade-umsatz-in-deutschland/ (accessed 6 June 2016).
Statista (2016b) *Umsatz mit Bio-Lebensmitteln in Deutschland in den Jahren 2000 bis 2015*. Available at: http://de.statista.com/statistik/daten/studie/4109/umfrage/bio-lebensmittel-umsatz-zeitreihe/ (accessed 6 June 2016).
Torjusen, H., Sangstad, L., O'Doherty Jensen, K. and Kjærnes, U. (2004) *European Consumers' Conceptions of Organic Food – A Review of Available Research*. Professional Report, no 4-2004. National Institute for Consumer Research, Oslo. Available at: http://orgprints.org/2490/ (accessed 30 June 2016).
Urena, F., Bernabéu, R. and Olmeda, M. (2008) Women, men and organic food: differences in their attitudes and willingness to pay. A Spanish case study. *International Journal of Consumer Studies* 32, 18–26.
Van Loo, E.J., Hoang Diem, M.N., Pieniak, S. and Verbeke, W. (2013) Consumer attitudes, knowledge, and consumption of organic yoghurt. *Journal of Dairy Science* 96, 2118–2129.
Vantomme, D., Geuens, M., de Houwer, J. and Pelsmacker, P. de (2006) Explicit and implicit determinants of fair-trade buying behaviour. *Advances in Consumer Research* 33(1), 699–703.
Verbeke, W. (2005) Agriculture and the food industry in the information age. *European Review of Agricultural Economics* 32, 347–368.
Vermeir, I. and Verbeke, W. (2006) Sustainable food consumption. Exploring the consumer 'attitude-behavioral intention' gap. *Journal of Agricultural and Environmental Ethics* 19(2), 169–194.
Wier, M. and Calverley, C. (2002) Market potential for organic foods in Europe. *British Food Journal* 104(1), 45–62.
Zagata, L. and Lostak, M. (2012) In goodness we trust. The role of trust and institutions underpinning trust in the organic food market. *Sociologia Ruralis* 52(4), 470–487.
Zander, K. and Zanoli, R. (2013) Consumer perception on organic farming. In: Sanders, J. (ed.) *Evaluation of the EU-Legislation on Organic Farming*. Final Report. Thünen Institute of Farm Economics, Braunschweig, Germany. Available at: http://ec.europa.eu/agriculture/evaluation/market-and-income-reports/organic-farming-2013_en.htm (accessed 03 November 2014).
Zander, K., Stolz, H. and Hamm, U. (2013) Promising ethical arguments for product differentiation in the organic food sector. A mixed methods research approach. *Appetite* 62, 133–142.
Zander, K., Padel, S. and Zanoli, R. (2015) EU organic logo and its perception by consumers. *British Food Journal* 117(5), 1506–1526.
Zanoli, R. and Naspetti, S. (2002) Consumer motivations in the purchase of organic food. A means-end approach. *British Food Journal* 104(8), 643–653.

7 Multiple Certifications and Consumer Purchase Decisions: a Case Study of Willingness to Pay for Coffee in Germany

Arnab K. Basu[1]*, Ulrike Grote[2], Robert Hicks[3] and Till Stellmacher[4]
[1]*Charles H. Dyson School of Applied Economics and Management, Cornell University, Ithaca, New York, USA;* [2]*Institute for Environmental Economics and World Trade, Leibniz University Hannover, Germany;* [3]*Department of Economics, College of William and Mary, Williamsburg, Virginia, USA;* [4]*Center for Development Research (ZEF), University of Bonn, Germany*

7.1 Introduction

Coffee produced in developing countries is sold under a variety of labels in the consumer markets of developed countries, the most prominent being Fair Trade. In addition to Fair Trade coffee, labels such as 'Organic', 'Shade Grown' and 'Wild Grown' coffee are also becoming popular amongst consumers in developed countries (Ponte, 2002). Labelling of coffee (as well as a variety of other products) under the Fair Trade umbrella aims to target two issues simultaneously: (i) to provide an income guarantee that acts as an insurance mechanism for poor farmers in developing countries who undertake production in a socially (for instance, no child labour in production) and environmentally friendly, sustainable way by following certain clearly defined standards; and (ii) to introduce market differentiation for a product that helps to correct for information distortion on the consumption side and allows consumers to reveal their true/higher willingness to pay (WTP). In effect, labelling provides for a market-based solution to the under-production and under-consumption of socially and environmentally friendly products while allowing for a higher income for farmers in developing countries. Thus, labelling can provide a superior alternative to direct production subsidies to farmers as an incentive to engage in environmentally friendly production techniques.

While a number of recent studies have focused on the supply side, assessing the impact of certification on producers' welfare and productivity and on consumers' WTP for specific labelled products, studies about consumers' perception and purchase decision making across multiple labels is non-existent. Consumer decision making in the context of multiple labels is an important issue, since the number of labels for a product such as coffee has substantially increased in the main consumer countries in the past two decades, creating not only more consumer choices and complexity, but also the possibility of label fatigue. Not only are labels such as Organic, Fair Trade, Wild Grown, 4 C, UTZ Certified, Rainforest Alliance, or Smithsonian Bird-friendly Shade Grown Coffee commonplace, but also double and triple certification such as Organic + Fair Trade, or Organic + Wild Grown + specific geographical

*Corresponding author; e-mail: ab362@cornell.edu

 Fair Trade and Organic Agriculture: A Winning Combination? (eds P. Parvathi, U. Grote and H. Waibel)

indication (such as Ethiopian Yirgacheffe or Jamaica Blue Mountain), are gaining prominence in the consumer markets. The study in this chapter fills the gap in consumers' choice and WTP for multiple certified coffee by focusing on three specific labels: 'Organic', 'Fair Trade' and 'Wild Grown'; and on the two large Arabica coffee production countries: Brazil and Ethiopia. In addition to identifying which certifications are relatively more valued by consumers, our study sheds light on whether older, more established labels like Fair Trade enjoy a market share advantage vis-à-vis newer labels like Wild Grown coffee.

7.1.1 Background literature

Organic certification started in the 1970s as a means of promoting and regulating eco-friendly agricultural practices. Since 1972, the International Foundation for Organic Agriculture (IFOAM) has been developing a global set of private and public regulations officially endorsed as organic, based on equivalence with the Common Objectives and Requirements of Organic Standards. All Organic labels indicate that the certified products have been produced and processed without chemical fertilizer, pesticides, fungicides, herbicides or genetically modified organisms (IFOAM, 2016). It is estimated that, in 2013, 260,000 t of Organic certified coffee were produced on 638,000 ha, representing 6.3% of the total global coffee area (Lernoud *et al.*, 2015) with 75% of the world's organic coffee being produced in Latin America alone (CBI Market Intelligence, 2016).

Fair Trade certification of coffee has increasingly gained worldwide importance (Nicholls and Opal, 2005; Petit, 2007; FLO, 2009; Grote *et al.*, 2009). Interestingly, coffee was the first product to be Fair Trade certified. In 1988, the first ever Fair Trade certified coffee was produced in Mexico and sold in The Netherlands. Today, coffee is the favourite product for Fair Trade consumers, accounting for 25% of all Fair Trade certified retail sales (FLO, 2016). The idea behind Fair Trade is to ensure 'fair' wages for small-scale producers in the developing countries (Raynolds *et al.*, 2007). Certification by Fair Trade International (FLO) targets poor farmers in developing countries operating on less than 3 ha of land without hired labour and guarantees them a price floor and other benefits like facilitating credit.[1] Fair Trade coffee guarantees not only that farmers and workers receive a fair price but also that it was produced without any child labour or forced labour, under healthy and safe working conditions, and with sustainable and environmentally friendly production methods. It provides a floor price to reduce farmers' vulnerability to coffee price fluctuations as well as a social premium that is targeted towards investments in schools, healthcare centres or other social infrastructure and facilities in the rural coffee community (FLO, 2015).

About 1.6 million farmers and workers, or an estimated 75% of all poor coffee farmers worldwide, are currently involved in Fair Trade (FLO, 2016). In recent years, Fair Trade operations have steadily increased: between 2011 and 2014, the area under production of Fair Trade certified coffee increased by 20% worldwide (Lernoud *et al.*, 2015). To put this production scale into perspective, in 2013 alone, 400,000 t of Fair Trade certified coffee was produced on more than 880,000 ha, constituting almost 9% of the global coffee area. Noteworthy here is that 34% of the globally produced Fair Trade certified coffee is also certified as organic, a fact that is exploited in our experiments. For consumers, mostly in the developed countries, the Fair Trade movement is generally perceived as a viable alternative to direct subsidies or aid and enhances the welfare of the poor producers in developing countries. This perception is evidenced in reality by worldwide consumer sales of the Fair Trade label that were estimated to be around $7.3 billion in 2013 (FLO, 2014).

Unlike the Fair Trade certification, Wild Grown coffee certification addresses environmental and biodiversity concerns by targeting preservation of natural habitats and ecosystems around coffee production areas. Wild Grown coffee is used as a label for Arabica coffee produced in its natural and original habitat, namely in the mountainous rainforests in southern and south-western Ethiopia. These forests, some of them still relatively undepleted, are the genetic cradle of *Coffea arabica*, today's most popular and highest-quality coffee species (Stellmacher, 2007; Stellmacher and Grote, 2011). The Wild Grown coffee label shows that this coffee grows wild in its natural habitat in forests with no or only little human interference. Wild coffee production in the rainforests of Ethiopia entails local farmers either simply picking wild coffee fruits inside these forests, or managing wild coffee stands by removing competing undergrowth vegetation and some canopy trees. It aims to prevent deforestation and preserve a unique ecosystem while also providing income support for local farmers in Ethiopia. It is interesting to note that the 'Wild Grown' coffee label is neither legally protected nor third-party accredited. Unlike Fair Trade labelled coffee, 'Wild Grown Coffee' from Ethiopia is an absolute niche product that has not been stocked on the shelves of larger retailers yet and, as a result, data about production, sales and market shares are not available.

To underscore the importance of the location of our study, Germany is the third biggest coffee-consuming country in the world, with 552 million kilograms in 2013. The per capita consumption in Germany is one of the highest worldwide, with 5.2 kg/person in 2013. Brazil is by far the largest country of origin for coffee consumed in Germany (Caffeine Informer, 2014). In 2014, the sales volume of Fair Trade certified coffee in Germany was 16,500 t. Two-thirds of the Fair Trade coffee sold in Germany is also Organic certified (TransFair, 2016). The sales of organic coffee in Germany were around 13,000 t in 2013, making Germany the largest organic coffee market in Europe and the second largest worldwide [(2)] (CBI Market Intelligence, 2016).

The literature on WTP for certified coffee generally shows that consumers are willing to pay a price premium but the magnitude of this premium depends on the label in question, the country of origin of the coffee (Howard and Allen, 2010) and the demographic profile of the consumers themselves (Van Loo *et al.*, 2015). For instance, Pelsmacker *et al.* (2006) found that, on average, Belgian consumers were willing to pay a 10% premium on the market price for products with a Fair Trade label. Rotaris and Danielis (2011) conducted a choice experiment and found that Italian households were willing to pay a premium of about €2.20 for a 250 g Fair Trade coffee packet. The WTP, however, varied significantly depending on age, gender, income and purchasing habits of the consumers. Loureiro and Lotade (2005) found the existence of a price premium for Fair Trade, Shade Grown and Organic coffee, though the premium for organic coffee was lower than that for the Fair Trade label. Yang *et al.* (2012) found that Chinese consumers were on average willing to pay 22% more for a medium cup of Fair Trade coffee than for a non-certified one. This was especially true for female consumers, consumers who made their own coffee and consumers who planned to increase their consumption in the following year. Finally, Hiscox *et al.* (2011) examined consumers' WTP for Fair Trade coffee using eBay auctions. By posting otherwise identical products on eBay, differences in winning auction prices indicated that consumers were willing to pay approximately 23% more for Fair Trade coffee.

Studies that have analysed the elasticity of demand and the retail pricing strategy for certified coffee include Andorfer and Liebe (2014), who showed that German consumers were likely to increase their Fair Trade coffee consumption in the face of a price reduction. Arnot *et al.* (2006) found that consumers of Fair Trade labelled coffee were less price sensitive and Grote (2009) found a significant premium for labelled coffee in Germany and noted that: (i) conventional coffee cost around €1.5/250g while retail prices for organic and Wild Grown coffees amounted to up to €8.50/250g; (ii) multiple uses of labels seemed to have an effect on the price level and price variation; and (iii) both conventional and organic coffees were always more expensive if the country or region of origin was mentioned.

A couple of studies have analysed how perceptions about the performance of labels affected consumers' WTP for coffee. Basu and Hicks (2008, 2016) conducted experiments that gave consumers additional information regarding the number of farmers participating in the Fair Trade programme and the revenue increases for these farmers due to the price guarantee of the programme. Their results showed that consumers' WTP exhibited an 'inverted-U' shape – the WTP rose as the number of participating farmers and revenue accruing to them increased, reached a maximum and declined after around 60% of poor farmers were included in a Fair Trade programme within a country.

Alongside studies that estimate consumers' WTP, elasticity of demand and the role of information on labels for certified coffees, a group of studies have explored whether: (i) consumers prefer a market-based solution to poverty reduction for farmers in developing countries through their WTP a premium for certified coffee or whether they prefer a direct intervention through subsidies; and (ii) whether consumers have a preference in the way the price premium is transferred to the recipient farmers – in-kind versus cash transfers. Koppel and Schulze (2013) used a revealed preference experiment in Germany to test whether consumers preferred a transfer through the purchase of Fair Trade coffee or a direct transfer through aid to poor farmers in developing countries. Evidence suggested that consumers had a strong preference for the indirect transfer mechanism through the purchase of Fair Trade coffee. Chiu *et al.* (2017) found that, on the one hand, consumers in the USA who preferred most of the premium to be an in-kind transfer to the recipients had a WTP of around 50% over standard coffee. On the other hand, the WTP of those who preferred most of the premium to be paid in cash amounted to around 40%, while those who were indifferent to how the premium was transferred to the recipient had a WTP close to 20%.

7.1.2 Motivation

This study contributes to the literature on consumers' WTP for certified coffee with the main objective of trying to understand consumer choices in the presence of: (i) multiple labels; and (ii) multiple certifications within each label type. The multiple labels we consider are Fair Trade, Wild Grown and Organic coffee while multiple certifications within each of these three main labels include permutations such as Fair Trade + Organic, or Organic + Wild Grown. Our choice of the three labels, Fair Trade, Wild Grown and Organic, is based on two factors. Firstly, the Fair Trade and Organic labels are older and well established in the German consumer market, while the Wild Grown label is relatively new. An interesting question here is whether the Wild Grown label being a late entrant has been able to penetrate the certified coffee market or whether the market is still dominated by the Fair Trade label. Secondly, the objectives of two of the labels, Fair Trade and Wild Grown, are quite different. While the Fair Trade label's main goal is to promote the welfare of poor producers in developing countries through a price guarantee with environmental sustainability as a secondary goal, the Wild Grown label's primary goal is to preserve an ecosystem through sustainable farming with farmers' incomes as the secondary goal. Thus, an interesting question here is to test whether consumers put more weight on social vis-à-vis environmental considerations in their purchase decisions, holding all other attributes of these two labels (prices and country of origin) constant. Finally, the use of multiple certifications allows us to test whether, and to what extent, the likelihood of preference for either the Fair Trade or Wild Grown coffee is driven by the additional attributes like Organic and country of origin.

7.2 Research Design and Methodology

7.2.1 Research design

We investigate the stated preferences of consumers over products having various label attributes such as Fair Trade, Wild Grown, Organic, country of origin and price. The choice experiment is designed to examine the sensitivity of how these various attributes impact the valuation of Fair Trade and Wild Grown coffee by German consumers. This methodology has been used for a wide variety of consumer products and has been shown to yield reliable information about market choices of respondents (e.g. Ben-Akiva and Morikawa, 1990; Adamowicz *et al.*, 1994). A major advantage of this methodology in our context is that it generates a large number of observations and structural estimation of preferences with fairly flexible functional forms. Blocked experimental design techniques were used to select the 12 sets of six questions for each experiment that maximizes the trade-offs of the coffee choice experiments.[(3)] The actual levels of the attributes chosen by the experimental design algorithm differ by question, block and experiment. Details on the attribute levels are presented in Table 7.1. We determined the price level in Table 7.1 by comparing prices in different sales outlets (discounter, specialty shops, etc.) where coffee is available. Conventional coffee cost around €1.50/250 g, while retail prices

Table 7.1. Choice sets, attributes and levels.

Attributes	Levels	Definitions
Price	€1.50 €3.50 €4.25 €4.50 €5.00	Price per 250 g packet of coffee based on comparison across various outlets. Conventional coffee costs around €1.50/250 g, retail prices for organic and wild coffees amount to up to €8.50/250 g
Country of origin	Not specified Brazil Ethiopia	Origin of the coffee beans
Certificate	Not specified Fair Trade Organic (Bio) Wild Grown	Information about each certification was presented at the beginning of the survey

for Organic and Wild Grown coffee amounted to up to €8.50/250 g. The most common countries of origin were Brazil and Ethiopia.

Each respondent was randomly assigned to one of the 12 blocks. Within each block a respondent received six sets of choices, with each set consisting of three coffee choices and an opt-out option. An example of a set of choices faced by a respondent is presented in Fig. 7.1. Respondents were given information about each certification scheme (Fair Trade, Wild Grown and Organic (Bio)) at the beginning of the survey and were explicitly instructed to assume that the product quality is identical across all three choices. In our hypothetical choice experiment, it is important for respondents to believe that the product quality and price are not positively correlated.[4] The experiments were conducted at the Leibniz University Hannover in January–April 2016 amongst undergraduate students, their friends and families. We had a total of 149 respondents with 2614 observed coffee choices.

7.2.2 Summary statistics

In the questionnaire we solicited information from the respondents about: (i) their social and environmental preferences; (ii) their awareness about existing Fair Trade, Wild Grown and Organic certifications in Germany; (iii) their preference for the certifying agency; and (iv) their educational and income profiles. The results are presented in Figs 7.2–7.5. To summarize, Fig. 7.2 shows that most respondents found it very important that farmers who grow coffee can achieve a higher standard of living; the opinions ranged from somewhat important to extremely important. Only a very small share of respondents indicated that this was not important to them, while there was general consensus among German respondents that coffee should be produced without child labour. Regarding environmental awareness, the survey revealed that it was important to the majority of respondents that coffee was produced using environmentally friendly and sustainable farming methods. There was also a high degree of consensus that coffee had to be free of chemical residues and should not be genetically modified. Interestingly, it was even more important to the respondents that coffee needed to be cultivated in a way that enhanced biodiversity – all of which pointed to consumers having a strong preference for a label that provided a public good.

Figure 7.3 captures respondents' awareness about the various labels of interest to us. Most consumers indicated that they knew quite well the characteristics of Organic coffee. A slightly lower but still quite high share of consumers was familiar with the characteristics of Fair Trade coffee. Familiarity with Wild Grown coffee was much less pronounced, which is not surprising as this is not a label widely available in the major sales outlets. Figure 7.4 shows that a majority of respondents indicated that a nongovernmental organization (NGO) should be in charge of the certification but many of the

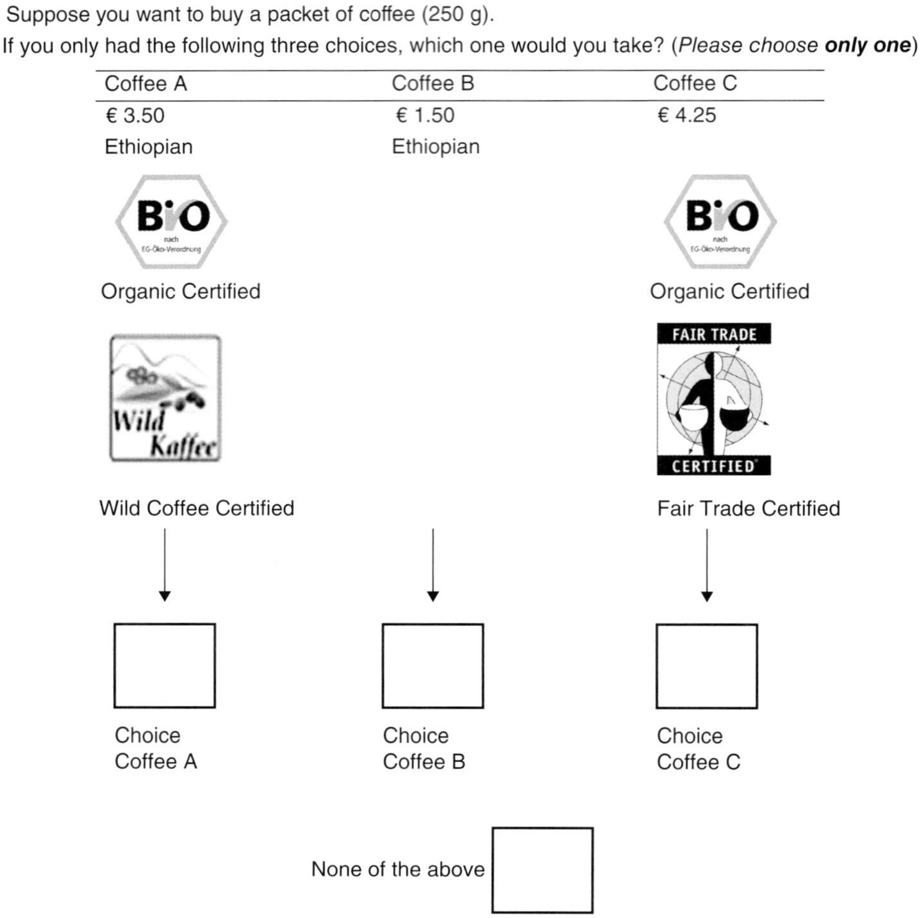

Fig. 7.1. Sample set of choices faced by a respondent.

German consumers also liked the idea of a German Government Office or the European Union being in charge of the certification. A smaller share of respondents saw the countries of origin as being in charge but retailers were not seen as being credible certifiers. Finally, Fig. 7.5 captures the characteristics of respondents. Most respondents were students, which is reflected by their age (between 16 and 25 years), academic degree (studying for a Bachelor's or Master's degree) and an average annual income of less than €5000.

7.2.3 Econometric methodology

Our experimental design allows for the identification of all the main and two-way interaction effects. For a coffee k presented to respondents, the utility is expressed as equation 7.1 at the bottom of the page.

where the attribute levels in the experimental design are 'turn on' parameters. For example, for coffees having no country of origin label, the term $(None_k == 1)$ would equal one, whereas $(Ethiopia_k == 1)$ and $(Brazil_k == 1)$ would both be

$$U_k = \beta_{price}P_k + \beta_{None}(None_k = 1) + \beta_{Ethiopia}(Ethiopia_k = 1) + \beta_{Brazil}(Brazil_k = 1) + \beta_{FT}(FT_k = 1) + \beta_{Organic}(Organic_k = 1) + \beta_{Wild}(Wild_k = 1) + \beta_{FT.BIO}(FT_k = 1)(BIO_k = 1) + \beta_{Wild.Organic}(Wild_k = 1)(Organic_k = 1) + \beta_{FT.Wild}(FT_k = 1)(Wild_k = 1) + \epsilon_k \quad (7.1)$$

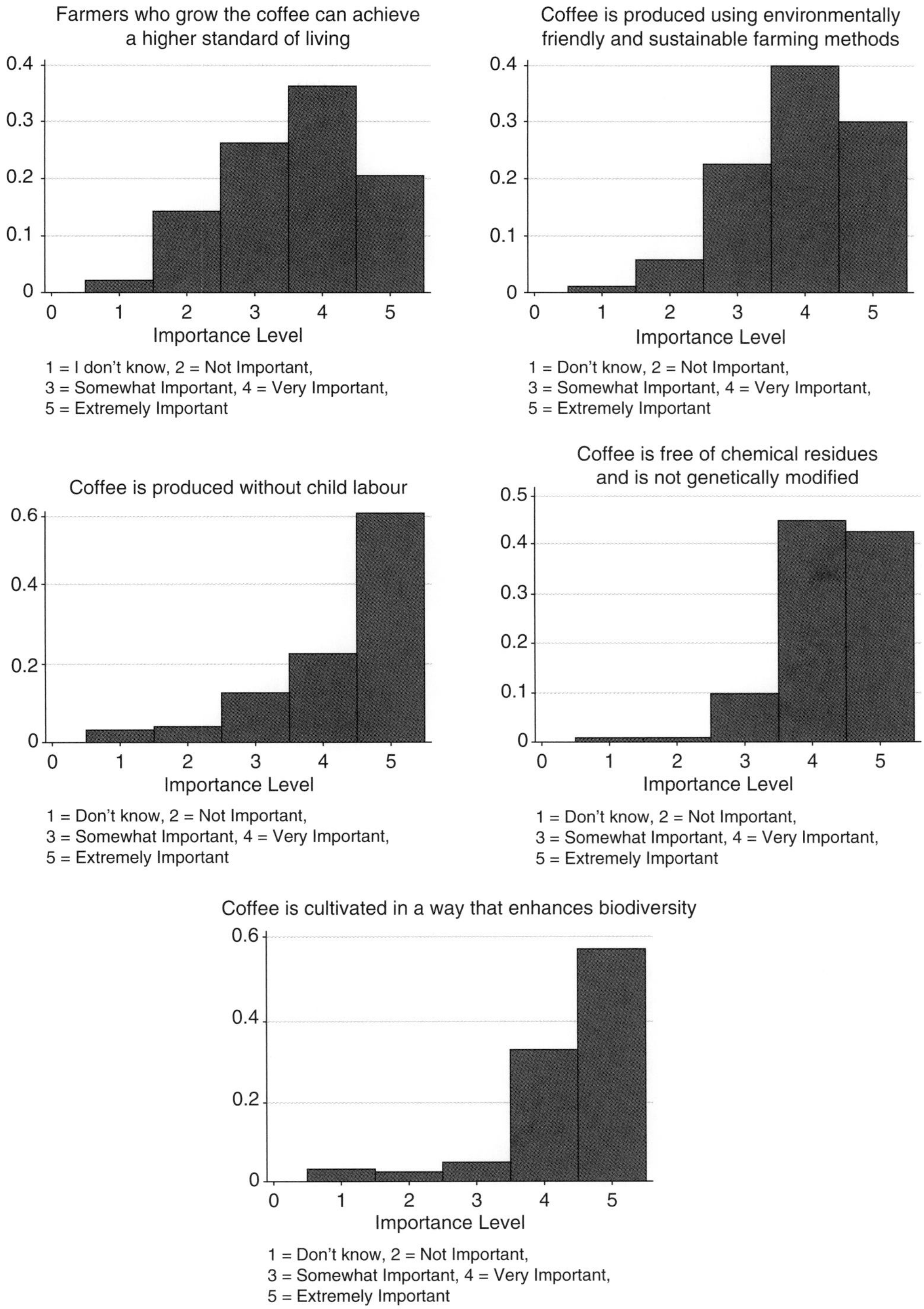

Fig. 7.2. Social and environmental awareness.

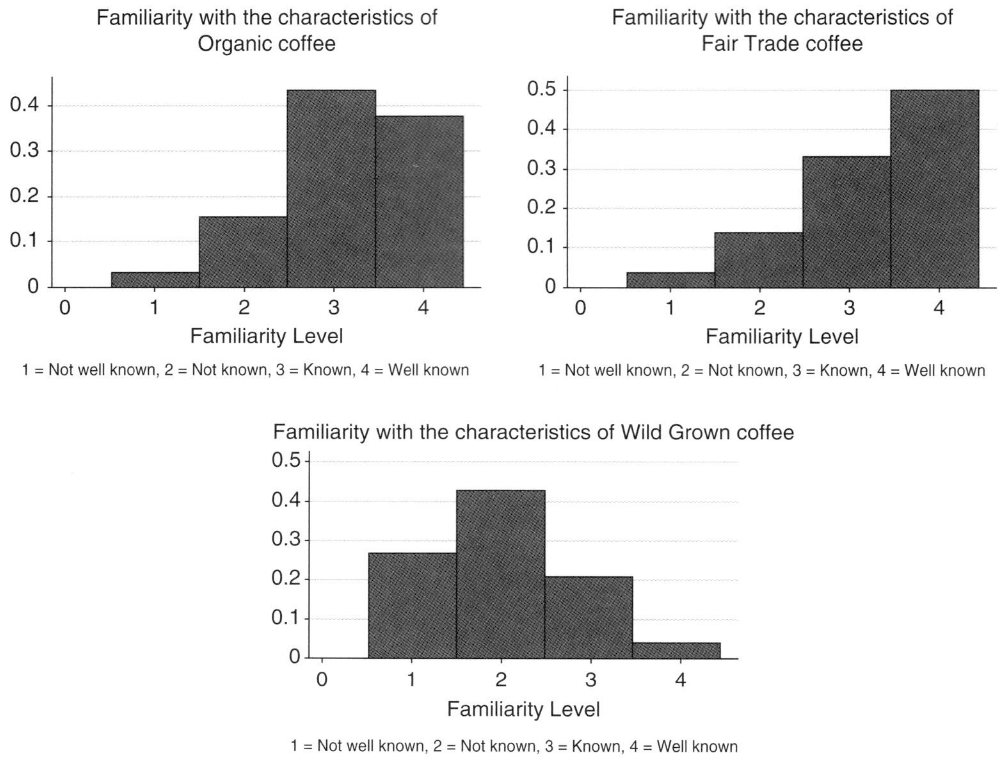

Fig. 7.3. Awareness of coffee characteristics.

zero. Unlike the country of origin attribute, it is possible for multiple labels to appear for a coffee presented to respondents for the Fair Trade (*FT*), Organic Label (*BIO*), and Wild Grown (*Wild*) attributes. This allows us to identify interaction effects such as $\beta_{FT,Organic}$. Below, we further outline the basis for our calculations for the WTP distributions for single and double certified coffees that are presented in Tables 7.2 and 7.3 and Figs 7.6 and 7.7 in Section 7.3 (Results).

Single certification:

Denoting 0 as the baseline unlabelled coffee, the indirect utility from consuming this baseline coffee is simply:

$$U\left(P^0,0\right)=\beta_p P^0+\beta_N+\epsilon_0 \tag{7.2}$$

We next compare the baseline indirect utility in equation 7.2 to the indirect utility of consuming the three single-labelled coffees – Fair Trade, Organic and Wild:

$$U\left(P^0,FT\right)=\beta_p P^0+\beta_N+\beta_{FT}+\epsilon_1 \tag{7.3}$$

$$U\left(P^0,Organic\right)=\beta_p P^0+\beta_N+\beta_{Organic}+\epsilon_2 \tag{7.4}$$

$$U\left(P^0,Wild\right)=\beta_p P^0+\beta_N+\beta_{Wild}+\epsilon_3 \tag{7.5}$$

Holding the coffee price and country of origin constant across all coffees, we can extract the expected WTP for coffee *i* by solving:

$$E\left[U\left(P^0+WTP^i,i\right)=U\left(P^0,i\right)\right] \tag{7.6}$$

Thus, for example the willingness to pay for Fair Trade labelled coffee (WTP^{FT}) is:

$$E\left[\begin{array}{r}\beta_p\left(P^0+WTP^{FT}\right)+\beta_N+\beta_{FT}+\epsilon_1\\ =\beta_p P^0+\beta_N+\epsilon_0\end{array}\right] \tag{7.7}$$

or:

$$E\left[\beta_p WTP^{FT}+\beta_{FT}\right]=0 \tag{7.8}$$

Equation 7.8 can then be simplified to: $E\left[WTP^{FT}\right]=\frac{-\beta_{FT}}{\beta_p}$. Thus, the expected WTPs for Bio and

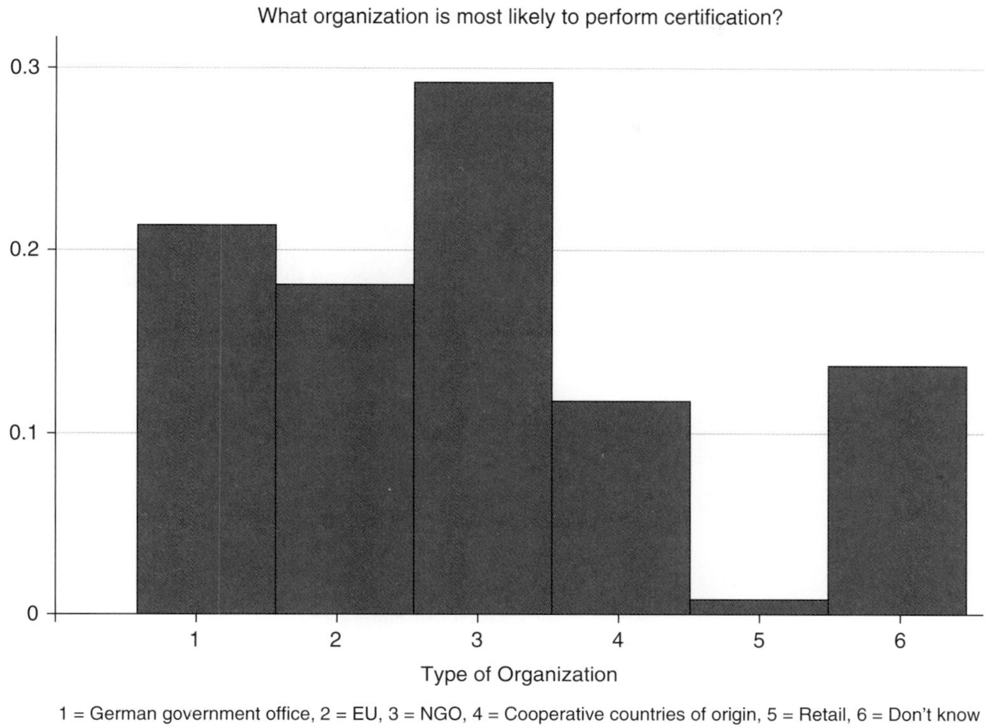

Fig. 7.4. Preference for certifying agency.

Wild coffees are, respectively, $E[WTP^{BIO}] = \frac{-\beta_{BIO}}{\beta_p}$ and $E[WTP^{Wild}] = \frac{-\beta_{Wild}}{\beta_p}$.

Double certification

For double certification where the labels FT + Organic or Wild Grown + Organic or FT + Wild Grown are combined, WTP calculations require modification of Eqns 7.3–7.8 as follows:

$$U(P^0, FT, Organic) = \beta_p P^0 + \beta_N + \beta_{FT} + \beta_{Organic} + \beta_{FT,Organic} + \epsilon_4 \quad (7.9)$$

$$U(P^0, Organic, Wild) = \beta_p P^0 + \beta_N + \beta_{Organic} + \beta_{Wild} + \beta_{Organic,Wild} + \epsilon_5 \quad (7.10)$$

$$U(P^0, FT, Wild) = \beta_p P^0 + \beta_N + \beta_{FT} + \beta_{Wild} + \beta_{FT,Wild} + \epsilon_6 \quad (7.11)$$

where $\beta_{i,j}$ (e.g. $\beta_{FT,BIO}$) is an interaction term if the coffee has double labels *i* and *j*. We can then solve for expected WTP for double-labelled coffees as compared with the baseline coffee (captured by Eqn 7.2). For example, the expected WTP for a coffee labelled as both Fair Trade and Bio would be:

$$E(WTP^{FT,Organic}) = \frac{-(\beta_{FT} + \beta_{Organic} + \beta_{FT,Organic})}{\beta_p} \quad (7.12)$$

Triple certification

For triple certified coffees such as FT + Organic + Wild, we can replicate the steps in Eqns 7.9–7.12 with an additional label, and the analogue of Eqn 7.12 would be as in equation 7.13 at the bottom of the page.

Consumer preference parameters for labelled coffees are estimated using data from the

$$E(WTP^{FT,Org.,Wild}) = \frac{-(\beta_{FT} + \beta_{Org.} + \beta_{Wild} + \beta_{Org.,FT} + \beta_{Wild,Org.} + \beta_{FT,Wild})}{\beta_p} \quad (7.13)$$

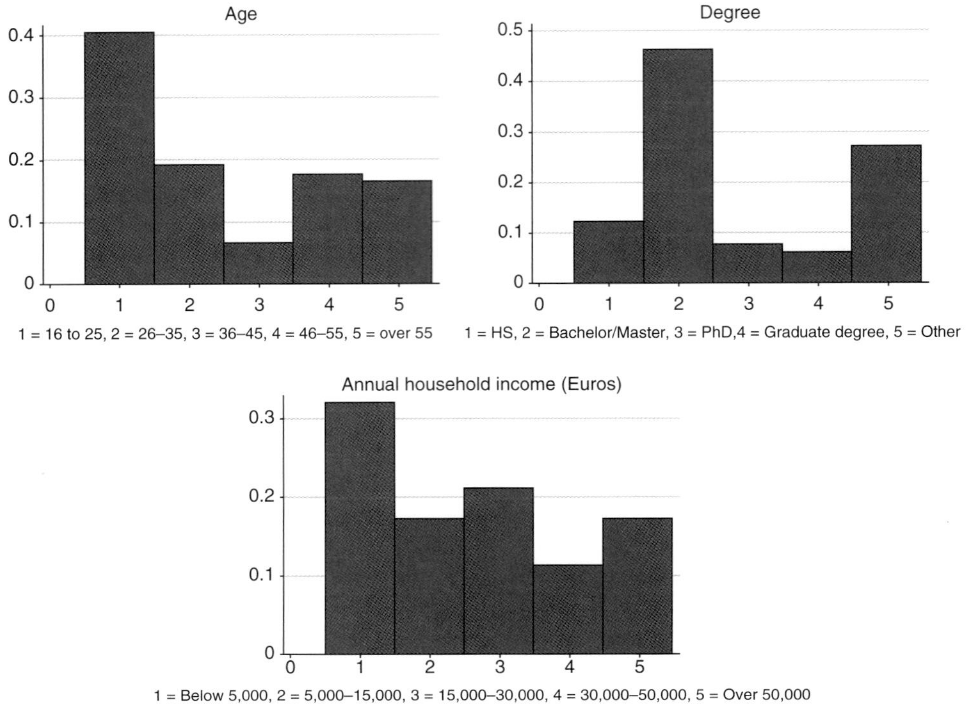

Fig. 7.5. Demographic profile.

experimental design outlined in Table 7.1 and depicted in Fig. 7.1. We implement a discrete choice model using a multinomial logit approach where alternatives are defined by each coffee alternative. Our estimation approach samples the posterior of the distribution of our parameter estimates by constructing a Monte Carlo Markov Chain, where the posterior is defined as $P(\beta | d, \boldsymbol{X}) \propto P(d | \beta, \boldsymbol{X}) P(\beta | \beta^0)$. $P(d|\beta,\boldsymbol{X})$ is the likelihood function for a multinomial logit model, $P(\beta | \beta^0)$ are our priors on the model parameters, $\boldsymbol{X}$ are the coffee attribute levels as given by the experimental design, and d indicates which coffee was chosen for each coffee choice in the data. In this study we assume flat priors (any real numbered parameter vector is equally likely), making our posterior $P(\beta | d, \boldsymbol{X}) \propto P(d | \beta, \boldsymbol{X})$.

Consequently, when constructing our Monte Carlo Markov Chain to sample from the posterior distribution of parameters, we are sampling exactly from the distribution of parameters that maximizes the likelihood. The primary advantage of our estimation approach versus classical Maximum Likelihood (since it uncovers the maximum likelihood estimates) is that we can simultaneously sample from the distribution of our willingness-to-pay measures while estimating the model parameters.[5]

7.3 Results

Table 7.2 contains our estimates about consumers' preferences for labelled coffee. Note that for each estimated variable we report the posterior mean and standard deviation (analogous to the maximum likelihood parameter estimate and standard error). While not reported, all P-values are significant at the 5% (and even 1%) level except for Wild Grown coffee, and the Organic and Wild Grown coffee interaction term. In addition, we report percentiles of our sampled parameter values including the median, minimum and maximum. Using the 2.5% and 97.5% percentiles, one can construct the 95% confidence interval for each parameter. For example, the 95% confidence interval for the Fair Trade parameter is [2.374, 3.172].

Table 7.2. Parameter estimates.

	Mean	Std Dev	Lower 2.5% CI	Upper 2.5% CI
β_{price}	–0.683	0.059	–0.798	–0.569
β_{Brazil}	1.477	0.201	1.087	1.876
$\beta_{Ethiopia}$	1.456	0.211	1.045	1.867
β_{None}	1.415	0.215	0.994	1.834
β_{Bio}	2.450	0.202	2.059	2.850
β_{FT}	2.771	0.205	2.374	3.172
β_{Wild}	–0.206	1.320	–2.856	2.086
$\beta_{FT, Bio}$	–1.853	0.278	–2.405	–1.314
$\beta_{FT, Wild}$	–0.637	0.284	–1.180	–0.071
$\beta_{Bio, Wild}$	0.685	1.327	–1.649	3.327

Table 7.3. Willingness-to-pay estimates.

	Mean	Std Dev	Lower 2.5% CI	Upper 2.5% CI
WTP^{FT}	4.070	0.262	3.567	4.586
WTP^{Bio}	3.601	0.306	3.010	4.209
WTP^{Wild}	–0.306	1.957	–4.212	3.116
$WTP^{Wild+FT}$	2.825	2.005	–1.121	6.483
WTP^{Bio+FT}	4.951	0.370	4.218	5.663
$WTP^{Bio+Wild}$	4.304	0.307	3.710	4.906
$WTP^{Bio+Wild+FT}$	4.715	0.312	4.124	5.340

The signs and significance levels of all one-way effects (except for Wild Grown coffee) accord with our expectations. The results show that (assuming all other coffee attributes are equal) consumers prefer cheaper coffee, coffee labelled with Fair Trade and Organic labels, and weakly prefer coffee having a labelled country of origin. For interaction terms, we see that coffees with multiple labels typically have a negative sign, indicating that the whole is less than the sum of its parts: consumers prefer additional labels but their utility is increasing at a decreasing rate. The only exception to this is the Organic and Wild Grown coffee interaction term, which is not significant.

We also report on the consumers' WTP for a number of possible combinations of labels versus a generic unlabelled coffee in Table 7.3. Recall that for label *i*, the mean willingness to pay is $[WTP^i] = \frac{-\beta_i}{\beta_p}$. For multiple labels, the WTP measure is given as:

$$(WTP^{i,j}) = \frac{-(\beta_i + \beta_j + \beta_{i,j})}{\beta_p}$$

Focusing on the single-labelled coffees (WTP^{FT}, $WTP^{Organic}$, WTP^{Wild}), results show that both the Fair Trade and Organic labels are very valuable to German consumers whereas we do not see evidence that they value Wild Grown labelled coffee. For coffees with multiple labels, we see that the Organic and Fair Trade labelled coffee is the most valuable coffee on average of any other label considered in this study, including the triple-labelled coffee ($WTP^{FT+Wild+Organic}$). There is some evidence of complementarity between the Organic and Wild Grown coffees (mean WTP of 4.30) since taken together the WTP is greater than the WTP for Organic. However, we do not see this complementarity between Wild Grown and Fair Trade coffees (mean of 2.84).

Figure 7.6 (for the single-labelled coffees) and Fig. 7.7 (for the multiple-labelled coffees) explore in more detail the posterior distributions of consumers' WTP for labelled coffees. We see that consumers have high valuation (and relatively low dispersion) for the Organic and Fair Trade coffees, whereas we see consumers do not have any preferences (statistically speaking) for Wild Grown coffee and are not willing to pay a positive price premium for this label. In Fig. 7.7, we again see some evidence of complementarity between coffees labelled as Organic and Wild Grown compared with coffees labelled as Fair Trade and Wild Grown. The latter has a very high degree of dispersion and we cannot say with confidence that the consumer values the coffee over the generic unlabelled one. We also see that the Fair Trade and Organic labelled coffees are the most valuable of the coffees considered here, even marginally more valuable than the triple-labelled coffee (Fair Trade, Organic, and Wild Grown). These results suggest that when a relatively unknown coffee attribute (Wild Grown coffee) is added to existing labelling schemes, the multiplicity of labels sometimes causes WTP to fall compared with coffees with fewer but established labelled attributes.

7.4 Conclusion

This study set out to explore consumer purchase decisions for certified coffee in a setting where the choice set included multiple labels and multiple certifications for each label. An important issue that we wanted to address was whether earlier certifications have an inherent

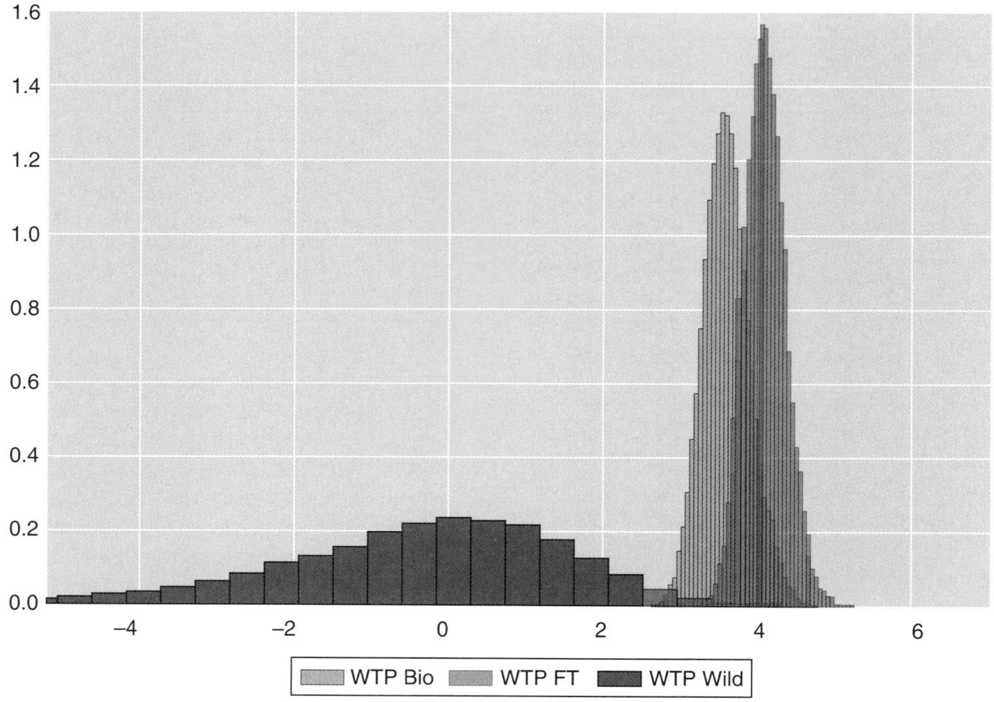

Fig. 7.6. Willingness to pay (WTP) for various single labels.

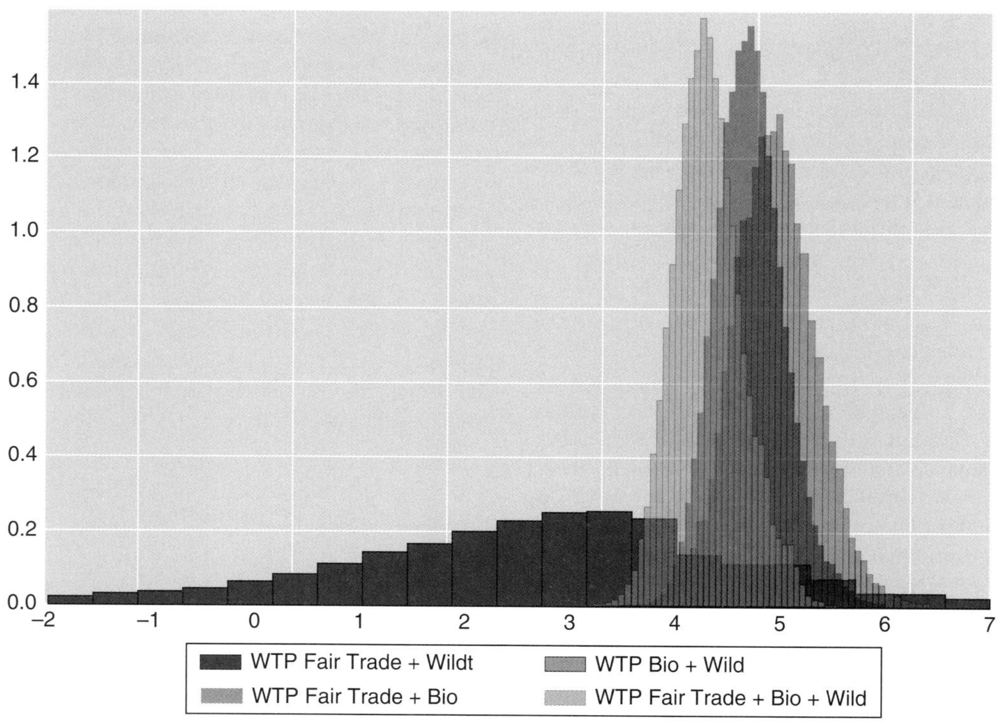

Fig. 7.7. Willingness to pay (WTP) for single and double labels.

advantage in the sense that more recent certifications are less likely to be chosen by consumers. Our results indicate that there is a discrepancy between consumers' responses with regards to their preference for environmental sustainability and biodiversity and their observed choices when faced with alternatives. Specifically, we see that while the majority of consumers feel that environmental sustainability and biodiversity is important, they tend to avoid choosing Wild Grown coffee when presented with the alternatives of Fair Trade and Organic coffee. Given that the Wild Grown label is a fairly new one (compared with Fair Trade and Organic labels), possible explanations of this observation could well be consistent with the following: (i) consumers are not aware of the exact benefits that the Wild Grown label imparts on the environment; (ii) Fair Trade and Organic labels are older and have an established market; and (iii) consumers when faced with a pure social label (Fair Trade) and a pure eco-label (Wild Grown) tend to favour the social label. These results imply that the target group should be identified and their WTP estimated prior to promoting a new label. It also suggests that social issues are important aspects to consider when investing in a new label. While the design of our study prevents identifying the precise reason as to why the Wild Grown label is not yet popular with German consumers, we hope that this study encourages further investigations into: (i) the issue of consumer decision making in the face of multiple certifications; and (ii) the possible market share advantage that older, established labels enjoy.

Notes

(1) In 2011 Fair Trade USA split from Fair Trade International (FLO), and while FLO's certification applies to 'small' farmers owning less than 3 ha of land, Fair Trade USA certification also includes farmers who own more than 3 ha and employ hired labour on their farms.

(2) In 2013, Germany's organic coffee imports accounted for around 19% of the world's total. Globally, Germany is second after the USA, which imported about 38% of the world's organic coffee. These figures do not, however, include organic coffee produced in Peru, which is not an ICO member but is by far the world's largest exporter of organic coffee (CBI Market Intelligence, 2016).

(3) In practice, this means maximizing D efficiency or the determinant $|X' X^{-1}|$.

(4) A major issue with field experiments involving actual tasting of coffee would lead to the problem of a possible positive correlation between price and unobserved product quality.

(5) We sample from the posterior 20,000 times with a burn-in of 2000. Using standard diagnostics we can confirm that our Markov Chain has converged (which is needed for statistical inference).

References

Adamowicz, W., Louviere, J. and Williams, M. (1994) Combining revealed and stated preference methods for valuing environmental amenities. *Journal of Environmental Economics and Management* 26(3), 271–292.

Andorfer, V.A. and Liebe, U. (2014) *Do Information, Price, or Morals Influence Ethical Consumption? A Natural Field Experiment and Customer Survey on the Purchase of Fair Trade Coffee.* Social Sciences Working Paper 6. Department of Social Sciences, University of Bern, Switzerland.

Arnot, C., Boxall, P.C. and Cash, S.B. (2006) Do ethical consumers care about price? A revealed preference analysis of fair trade coffee purchases. *Canadian Journal of Agricultural Economics* 54(4), 555–565.

Basu, A.K. and Hicks, R.L. (2008) Label performance and willingness to pay for fair trade coffee: a cross-national perspective. *International Journal of Consumer Studies* 32, 470–478.

Basu, A.K. and Hicks, R.L (2016) *Poverty Alleviation through Social Labeling Programs? Information Valuation and Willingness to Pay for Fair Trade Coffee*. Mimeo, Dyson School of Applied Economics and Management, Cornell University, Ithaca, New York.

Ben-Akiva, M. and Morikawa, T. (1990) Estimation of switching models from revealed preferences and stated intentions. *Transportation Research* 24A(6), 485–495.

Caffeine Informer (2014) *Caffeine (Coffee) Consumption by Country*. Available at: https://www.caffeineinformer.com/caffeine-what-the-world-drinks (accessed 6 July 2017).
CBI Market Intelligence (2016) *CBI Product Factsheet: Organic Coffee in Germany*. CBI Ministry of Foreign Affairs. The Hague. Available at: www.cbi.eu/sites/default/files/product-factsheet-germany-organic-coffee-2016.pdf (accessed 20 July 2016).
Chiu, L.J.V., Gómez, M.I., Liaukonyte, J. and Kaiser, H.M. (2017) Socially responsible products: what motivates consumers to pay a premium? (Paper presented at 2016 Allied Social Sciences Association (ASSA) Annual Meeting, 3–5 January 2016, San Francisco, California.) *Applied Economics* 49 (19), 1833–1846.
FLO (2009) *Fair Trade Leading the Way –Annual Report 2008–09*. Fair Trade Labelling Organizations International, Bonn, Germany.
FLO (2014) *Strong Producers, Strong Future. Annual Report 2013–14*. Fairtrade International, Bonn, Germany. Available at: http://www.fairtrade.org.uk/en/media-centre/news/september-2014/fairtrade-delivering-change (accessed 6 July 2017).
FLO (2015) *The History of Fair Trade*. Available at: http://www.Fair Trade.net/history-of-Fair Trade.html (accessed 27 February 2016).
FLO (2016) Focus on Fairtrade products: Coffee. In: *Monitoring the Scope and Benefits of Fairtrade*, 7th edn. Fairtrade International, Bonn, pp. 72–79. Available at: www.fairtrade.net/products/coffee.html (accessed 25 July 2016).
Grote, U. (2009) Environmental labeling, protected geographical indications, and the interests of developing countries. *The Estey Centre Journal of International Law and Trade Policy* 10(1), 94–110.
Grote, U., Basu, A.K. and Chau, N.H. (eds) (2009) *New Frontiers in Environmental and Social Labeling*. Springer/Physica, Heidelberg, Germany.
Hiscox, M.J., Broukhim, M. and Lewis, C.S. (2011) Consumer Demand for Fair Trade: New Evidence from a Field Experiment Using eBay Auctions of Fresh Roasted Coffee. Unpublished Manuscript. Harvard University, Cambridge, Massachusetts. Available at: https://scholar.harvard.edu/files/hiscox/files/consumerdemandfairlaborstandardsevidencecoffee.pdf (accessed 6 July 2017).
Howard, P.H. and Allen, P. (2010) Beyond organic and fair trade? An analysis of ecolabel preferences in the United States. *Rural Sociology* 75(2), 244–269.
IFOAM (2016) *Family of Standards*. IFOAM Organics International, Bonn, Germany. Available at: http://www.ifoam.bio/en/search?find=family+of+standards (accessed 20 July 2016).
Kaffeeverband (2016) *Deutscher Kaffeemarkt*. Available at: http://www.kaffeeverband.de/presse/zahlen-daten-fakten/deutscher-markt (accessed 21 July 2016).
Koppel, H. and Schulze, G. (2013) The importance of the indirect transfer mechanism for consumer willingness to pay for fair trade products – evidence from a natural field experiment. *Journal of Consumer Policy* 36(4), 369–387.
Lernoud, J., Potts, J., Sampson, G., Voora, V., Willer, H. and Wozniak, J. (2015) *The State of Sustainable Markets. Statistics and Emerging Trends 2015*. International Trade Centre (ITC), Geneva, Switzerland.
Loureiro, M.L. and Lotade, J. (2005) Do fair trade and eco-labels in coffee wake up the consumer conscience? *Ecological Economics* 53(1), 129–138.
Nicholls, A. and Opal, C. (2005) *Fair Trade: Market-driven Ethical Consumption*. Sage Publications, London.
Pelsmacker, P. de, Janssen, W., Sterckx, E. and Mielants, C. (2006) Fair trade beliefs, attitude and buying behaviour of Belgian consumers. *International Journal of Voluntary Sector Marketing* 11(2), 125–138.
Petit, N. (2007) Ethiopia's coffee sector: a bitter or better future? *Journal of Agrarian Change* 7(2), 225–263.
Ponte, S. (2002) The 'Latte Revolution'? Regulation, markets and consumption in the global coffee chain. *World Development* 30(7), 1099–1122.
Raynolds, L.T., Murray, D. and Heller, A. (2007) Regulating sustainability in the coffee sector: a comparative analysis of third-party environmental and social certification initiatives. *Agriculture and Human Values* 24, 147–163.
Rotaris, L. and Danielis, R. (2011) Willingness to pay for fair trade coffee: a conjoint analysis experiment with Italian consumers. *Journal of Agricultural & Food Industrial Organization* 9(1), ISSN (Online) 1542-0485. doi: 10.2202/1542-0485.1305, June 2011.
Stellmacher, T. (2007) *Governing the Ethiopian Coffee Forests*. Shaker Publishing, Aachen, Germany.
Stellmacher, T. and Grote, U. (2011) Forest Coffee Certification in Ethiopia: Economic Boon or Ecological Bane? ZEF Working Paper 76. Zentrum für Entwicklungsforschung (Center for Development Research), University of Bonn, Germany.

TransFair (2016) *Kaffee Fact Sheet. Zahlen und Fakten*. TransFair, Köln. Available at: www.fairtrade-deutschland.de/fileadmin/DE/mediathek/pdf/fairtrade_kaffee_factsheet.pdf (accessed 25 July 2016).
Van Loo, E.J., Caputo, V., Nayga, R.M., Seo, H.-S., Zhang, B. and Verbeke, W. (2015) Sustainability labels on coffee: consumer preferences, willingness-to-pay and visual attention to attributes. *Ecological Economics* 118(C), 215–225.
Yang, S.-H., Hu, W., Mupandawana, M. and Liu, Y. (2012) Consumer willingness to pay for fair trade coffee: a Chinese case study. *Journal of Agricultural and Applied Economics* 44(1), 21–34.

8 Consumers' Information Search and Preferences for Fair Trade Coffee: a Case Study from Germany

Rosa Schleenbecker[1]*, Katrin Zander[2] and Ulrich Hamm[1]
[1]*Department of Agricultural and Food Marketing, University of Kassel, Germany;*
[2]*Thünen Institute of Market Analysis, Federal Research Institute for Rural Areas, Forestry and Fisheries, Braunschweig, Germany*

8.1 Introduction

The aim of this case study is to analyse consumers' preferences as reflected in their information search when shopping for Fair Trade products. Information and its credibility play a crucial role since Fair Trade and organic products are, as described in Chapter 6, so-called credence goods (e.g. Padel and Foster, 2005; Janssen and Hamm, 2012). Information provided by food suppliers has to match consumers' information needs in order to reduce the feeling of an information overload (Hwang and Lin, 1999; Verbeke, 2005). Knowing about consumers' knowledge, motives and information search behaviour gives marketers the basis for designing effective communication measures.

The study examines the preferences of Fair Trade consumers for different attributes of Fair Trade products. Therefore, consumers' information search behaviour prior to the purchase decision is analysed. The study summarizes the state of knowledge amongst Fair Trade consumers in Germany and takes a look at consumers' motivations for Fair Trade purchase, the role of labelling and the importance of organic product features. Based on the results of this study, marketers will be able to design Fair Trade products according to consumers' preferences and their information search behaviour in a targeted manner. Coffee is used as an exemplary product since it is a well known Fair Trade product with the biggest market share amongst Fair Trade products in Germany.

8.2 Methods, Research Design and Sample Description

A mixed-method approach combining qualitative and quantitative research was chosen. Focus group discussions (FGDs) were performed in order to explore the field, followed by quantitative purchase experiments using the Information Display Matrix (IDM). Three FGDs were conducted in order to get insights into Fair Trade consumers' information needs as well as to understand their level of knowledge on the products and the underlying standards. Based on the results of the FGDs, a quantitative study was conducted, including IDM experiments with 389 Fair Trade consumers in Germany. The IDM is a process tracing method that helps to identify information search and decision behaviour.

*Corresponding author; e-mail: rosa.schleenbecker@gmx.de

The IDM experiments were combined with a written questionnaire; both were carried out on laptop computers (Schleenbecker and Hamm, 2015).

8.2.1 Focus group discussions

In market research, FGDs are used in order to explore opinions, behavioural patterns and attitudes as well as underlying consumer motives. The goal of FGDs is to generate insights into relevant topics on which information is scarce (Lamnek, 2005; Halkier, 2010). Consequently, they are especially used in order to explore new fields of research. Usually more than one discussion round is held in order to collect the various points of view on a topic (Bryman, 2008).

Data was analysed with qualitative content analysis based on the approaches of Gläser and Laudel (2006) and Mayring (2008). During the qualitative content analysis, a text different from the original text is created which only contains the information that is relevant in order to analyse the research questions. Due to predefined interpretation steps, the qualitative content analysis is traceable for others and intersubjectively verifiable. The aim of qualitative content analysis is to analyse text material by extracting text elements that are relevant for answering the research questions (Mayring, 2008).

In this study, three FGDs were conducted with a total of 27 participants. They were carried out in three German cities in the federal states of Hesse and Lower Saxony in autumn 2010. The participants were approached randomly in public places and were screened based on the requirement that they consumed Fair Trade products at least occasionally. The potential participants were not only asked if they knew and bought Fair Trade products, but they also had to be able to describe how they recognized these products in order to make sure that they really were Fair Trade consumers. The groups mainly consisted of occasional buyers of Fair Trade products. Many participants stated that they bought at least three different Fair Trade products from time to time or more often. About half (15) of the participants were female, and two-thirds (18) were between 18 and 44 years old. Each FGD lasted about 1.5 h. Subsequent to the discussions, the participants received a expense allowance of €20. In all three groups, the same discussion guidelines were used. According to Lamnek (2005), the most important questions were listed to structure the focus group's discussions upon them. The guidelines served as a discussion basis; all questions were discussed, until they seemed to be redundant. Table 8.1 displays the three topics that were discussed as well as each section's core issues.

8.2.2 Information Display Matrix

Complementary to the qualitative approach, Information Display Matrix experiments were conducted. Participants are offered different products and are asked to make a purchase decision. Prior to the purchase decision, participants can retrieve as much information as they need to make their decision, just as in a real shopping environment. Nowadays, the IDM is usually computer based.

The matrix is shaped by the different products (columns) and product attributes (rows). The resulting fields contain the actual product

Table 8.1. Core issues of the group discussions.

Topics	Core issues
Perception of information on Fair Trade and information search	• How intensively do consumers search for information on Fair Trade products? • Is the existing product information sufficient for consumers? • Is there a need for further information?
Labelling	• What are consumers' expectations towards Fair Trade labelling? • What relevance has the display of an organic label on Fair Trade products?
Knowledge, relevance and purchasing motives	• What is the state of knowledge of consumers on Fair Trade? • What are consumers' purchasing motives? • Which Fair Trade attributes are relevant for consumers?

information. The specific information can be accessed by clicking on the fields with the computer mouse (Fig. 8.1).

The most important advantage of the computer-based IDM is that the bias between actual and indicated behaviour (Aschemann-Witzel and Hamm, 2011; Zander and Hamm, 2012) is minimized compared with direct inquiries, due to the fact that participants interact with a computer instead of a person (cf. Bailey, 1994). Therefore, social desirability behaviour is minimized, which is especially important when ethical consumption is to be examined.

The IDM is a suitable method for identifying search and decision behaviour, since consumers are used to information being presented diagrammatically as in online shops and also in product test magazines, as prior research has shown (Aschemann-Witzel and Hamm, 2011; Zander and Hamm, 2012).

The whole information search process was recorded by the computer. As a result, the decision-making process including tracking of information search, evaluation of alternatives and the purchase decision can be traced, which means that the amount, content and sequence of information search as well as the final purchase decision can be analysed.

For this case study a survey was conducted in five retail stores in the federal states of Hesse and Lower Saxony in Germany where Fair Trade coffee was available. Due to being conducted in a real shopping environment, the negative implications of artificial experimental situations could be minimized while still having the advantage of controlled experiment conditions. Different store types were selected in order to avoid a bias in the sample with regard to shopping places: two of the retail stores were conventional supermarkets, one was an organic food shop and two were worldshops. Participants were chosen at random and had to be purchasers of Fair Trade products on at least an occasional basis.

Participants could virtually choose from nine different Fair Trade coffees that were equipped with different product attributes, which were Fair Trade specifics, production method (organic or conventional), price and Fair Trade labelling. The tested Fair Trade specifics were: protection of children (such as tackling the problem of child labour); producer income (e.g. a price that secures standard of living for producers); trade relations (e.g. guaranteed purchase of harvest); and social projects (e.g. the establishment of a health centre). The attributes were selected based on the results of the FGD, the literature review and information found on Fair Trade

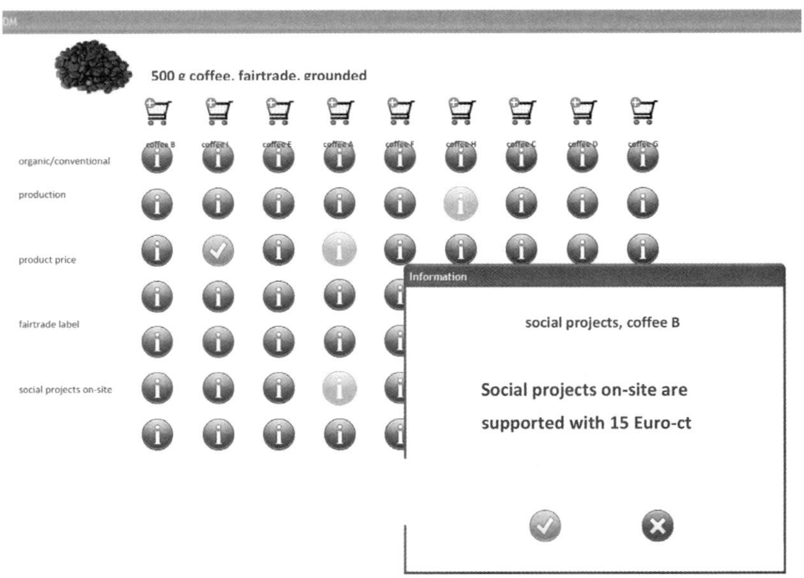

Fig. 8.1. Screenshot of the Information Display Matrix (IDM) with opened field.

products. The attributes were combined with the different products and shaped the matrix (Fig. 8.1). Each (initially hidden) field of the matrix contained the respective value of the product attributes or information on Fair Trade specifics.

The information on the Fair Trade attributes varied in three ways: information was presented in either a precise or imprecise manner, or the field contained no information. Options were: 'support of social projects on-site' (imprecise) and 'social projects on-site are supported with 15 Euro-cent [€0.15] of the retail price' (precise). Regarding the production method, products were either equipped with an organic label or no information was given (= conventional product). The Fair Trade labelling varied, too. Products were labelled either with the best-known TransFair label or with a faked label ('Fairvision') (Table 8.2). Participants were given these options in order to better understand label awareness.

In total, 63 different combinations of attributes were possible. Since not all of the theoretically possible attribute combinations could be tested, 18 different products were selected and presented in two different IDM designs (nine product options to choose from in each IDM design). They were tested with two randomly chosen groups of participants. A schematic overview over the product design is given in Table 8.2. Each IDM design contained a product that only had information on the Fair Trade label (either TransFair or Fairvision) and the price (see product I in Table 8.2.) The other displayed products were equipped with information on the production method (either the national organic logo or no information) and additional information on different Fair Trade aspects of the product. These were presented in either a precise or an imprecise manner. Attributes and products were arranged randomly in order to avoid a bias due to common reading habits from the upper left corner to the right.

Prior to the experiment, the task was explained to the participants. They were also informed that they had to choose and purchase one of the displayed coffees and that this purchase was binding. The task was supported by a monetary incentive of €5. Participants were allowed to access as much information as they liked and in the order they preferred by clicking on a field, which opened subsequently and contained the specified information on the attributes (see Fig. 8.1). This allowed tracing of the amount, content and strategy of information uptake with the help of a computer program. Fields could be opened repeatedly, but only one field could be opened at a time. Participants had the option of marking interesting attribute specifications. At the end of the process, participants had to put the product of their choice into a virtual shopping cart, as is the case in online shops.

The whole IDM experiment was embedded into a written questionnaire that elicited, inter alia, consumers' motivations to purchase Fair Trade products as well as socio-demographic data.

8.3 Results and Discussion

In the following subsections, the results of both research steps FGD and IDM are presented and discussed jointly. Firstly, consumers' state of knowledge on Fair Trade products is highlighted in Section 8.3.1. Purchase motivations are presented in Section 8.3.2. The specific information on Fair Trade in which consumers are interested is presented in Section 8.3.3, together with a discussion of the relevant product attributes which go beyond the Fair Trade and labelling aspects. This is followed in Section 8.3.4 by the extent of information search, the preferred information and how consumers search for information.

8.3.1 Consumers' knowledge on Fair Trade products

Recognition of Fair Trade products is the first important step in a consumer's purchase decision. The FGD showed that most consumers became aware of Fair Trade quality via 'the label'. Two participants emphasized that they would not have been able to recognize a Fair Trade product other than by the label: 'One doesn't have any other evidence but the label, isn't it?' Mostly, participants did not specify which 'label' they meant; only sporadically was 'the label' specified as the 'TransFair label', 'official label', 'FLO' or a German discounter Fair Trade label ('Lidl label').

Table 8.2. Overview over the schematic product design.

	Product A	Product B	Product C	Product D	Product E	Product F	Product G	Product H	Product I
Price	€5.99	€6.49	€5.99	€6.49	€5.49	€5.99	€5.49	€5.99	€4.99
Production method	BIO nach EG-Öko-Verordnung	BIO nach EG-Öko-Verordnung	BIO nach EG-Öko-Verordnung	BIO nach EG-Öko-Verordnung	–	–	–	–	–
Information on Fair Trade	imprecise	precise	imprecise	precise	imprecise	precise	imprecise	imprecise	–
Fair Trade label	FAIRTRADE	FAIRTRADE	FAIRVISION	FAIRVISION	FAIRTRADE	FAIRTRADE	FAIRVISION	FAIRVISION	FAIRTRADE

Some consumers recognized Fair Trade products via the German importer 'Gepa'. Individual participants explained that they recognized Fair Trade products via 'the headline', 'something with Fair Trade', or the description 'Fair Trade', a longer information text or the country of origin of the product.

The importance of the label for product recognition is supported by the results of the quantitative study, the IDM. Here, participants' first mouse clicks were mostly on the label, which underlines the importance of this attribute.

Knowledge of labels

The FGD investigated the state of knowledge on Fair Trade labels. Without directly inquiring which specific Fair Trade labels were known to the participants, it is interesting that most of the consumers just talked about 'the logo'. 'When I think of the labelling, only this logo where "Fair Trade" is written on it comes into my mind.' Participants also mentioned the 'best-known' or 'official' label, supposedly referring to the FLO label. Furthermore, several participants tried to describe the design of the label, working on basic principles, such as 'black and green, and kind of a water drop'.

Participants differentiated between several labels. The Fair Trade label of the German discount chain Lidl ('Fairglobe') was mentioned, but it was also indicated that different labels existed without the participant being able to name them. 'I think several logos do exist. This one was the best known, but there already exists kind of a logo-jungle.'

The results of the IDM experiments revealed that many participants purchased products with the fictional Fair Trade label. This share was 25% and may serve as an indicator that participants had difficulty in discerning between a real label and a fake label. This result underlines a rather poor level of knowledge on the labelling.

Production standards

Consumers' knowledge on production standards of Fair Trade commodities was mostly general and unspecific. For most participants, Fair Trade implies 'justice': 'For me, the justice actually ranks first, because so much in the world simply is totally unjust.' Further aspects of Fair Trade that were mentioned by many participants were living and working conditions of the producers, which should be 'fair', 'better' or 'humane'. 'Fair Trade producers' should be able to secure their families' living: 'I somewhat have the idea that the ... family which is involved, has somehow enough for lunch.' Also the payment of producers was often addressed and was expected to be 'fair', 'better' or 'appropriate': 'What is important is that producers are getting paid fair prices.' Banning child labour in production and ensuring school access for children were also referred to frequently. Some participants addressed self-determination of producers. Self-determination as regards to working time was discussed, as well as a lower dependency on 'landlords and wholesale', on 'international commodity markets' and on 'world trade prices'. Independence of producers was also understood as having the possibility to farm for the purpose of self-sufficiency. Guaranteed minimum prices as well as purchase quantities for producers and also pre-financing and micro credits were discussed.

These results are in line with earlier research. Wright and Heaton (2006) showed that hardly any consumers knew what Fair Trade means. These results cannot completely be confirmed since the participants in this study were by all means informed about the main features of Fair Trade. However, the participants in the present study were consumers of Fair Trade products, not consumers in general. The detected level of information regarding Fair Trade foods approximately coincides with the research of Beccheti and Rosati (2005), who found that more than 50% of the participants were informed about some, though not all, of the main features of Fair Trade. The present study also discovered that the majority of the participants was informed about general principles of Fair Trade, such as the payment of a 'fair' price and better working conditions. As in the study of Beccheti and Rosati (2005), only a few consumers were familiar with the more specific arguments of Fair Trade, such as pre-financing of production and price stability.

Results of a study by Daniel *et al.* (2010) showed that consumers misleadingly associated with Fair Trade quality other aspects that are actually not characteristics of Fair Trade, such as health and environmental friendliness. This can be confirmed by the present study, since participants generally related Fair Trade products with organic product quality. Daniel *et al.* (2010)

furthermore showed that Fair Trade is linked with more justice in world trade and support for the disadvantaged, as was the case with the participants of the FGDs. Individual participants, though, remarked that they perceived Fair Trade as a relationship on an equal footing with the producers, not as support by the privileged.

Organic farming and Fair Trade

Participants were not sure about the relationship between organic farming and Fair Trade. This topic was only marginally discussed. Some participants aligned Fair Trade with organic origin; others were unsure or became unsure whether there is a direct relationship. Some participants knew that some but not necessarily all Fair Trade products are certified as organic: 'Well, in the world shop they have chocolate ..., some organic, some not, and I think it was both Fair Trade.' In a study on Fair Trade and organic chocolate, Russeau (2015) also found that consumers were confused about the two concepts.

8.3.2 Purchase motives

In the FGD, five main purchase motivations were found. Some of the consumers stated that they would buy Fair Trade products for the sake of a positive feeling or in order to give themselves a good conscience: 'The purchase with a simply good feeling, indeed.' Other consumers bought Fair Trade products in order to reward themselves or others: 'Chocolate is really something that one gives away, or where you say: for a special occasion, I indulge myself in it.'

A smaller group of participants bought Fair Trade products because they aligned them with a better product quality and better taste: 'Well, for me it is all about quality, if the quality is really given. If it is better than a conventional product ...' Next to these hedonistic purchase motives, two altruistic purchase motives of justice and support of producer were mentioned: 'A little bit more justice, this is what it is about for me' and 'Well, that one supports the people, this I believe to be among the most important things.'

Some participants explained that the prevalent purchase motive and even the decision depended on the situation: 'I couldn't say what is most important. It totally depends on how I feel.' 'One also needs to be in the mood, and one needs to have the feeling that one wants to do something good, and when I go shopping then, then this comes together, I decide for this Fair Trade thing, not dogmatically, fundamentally, but in this very moment.'

The motives as identified in the FGD were, amongst others, quantitatively tested in the questionnaire that accompanied the IDM experiments (Table 8.3). The averagely highest approval could be detected for the statement: 'I buy Fair Trade products in order to make a contribution to more equity in the world.' The second reason to buy Fair Trade products was to support the producers and the third reason was for the good feeling.

These results are in line with earlier research of Tagbata and Sirieix (2008) as well as of Ferran and Grunert (2007), according to which consumers also express, in addition to altruistic motives, the hedonic motive for better product quality.

8.3.3 Relevant product attributes and information content

Due to the qualitative character of the FGD, no clear ranking of different attributes with respect

Table 8.3. Purchase motives.

I buy Fair Trade products (*n* = 389)	Mean value	Standard deviation
... in order to make a contribution to more equity in the world	4.61	0.76
... because I want to support the producers	4.47	0.89
... because the purchase makes me feel food	4.14	0.93
... because I think that the products have a good quality	4.11	0.80
... in order to indulge myself with something	3.35	1.16
... in order to reward someone	3.24	1.12

Scale: 1 = 'I don't agree at all', to 5 'I totally agree'

to their relevance for purchase decisions could be performed. Instead, the variety of different motives became obvious. Nevertheless, some motives were more prevalent in the discussions than others. In most cases, various product attributes had a high priority for the purchase decision. In addition to Fair Trade, for many participants the organic quality was rated high: 'First and foremost, I would take care that it is organic.' Often, it was emphasized that the organic attribute was most important, together with the quality or the taste of the product: 'Yes, taste and quality, ... if it is organic, is also crucial.' As far as addressed in the FGD, the participants ranked the organic quality higher than the Fair Trade quality. 'Well normally, I rather take care that the products are organic than from Fair Trade.' For other participants, the price played a crucial role for the purchase decision: 'I am also a regular coffee drinker, and I think, if I would buy Fair Trade coffee, I would become poor over it. I can't afford that.' For consumers who stated that they cared about the price, other product attributes were less relevant and Fair Trade products were mainly purchased for 'special moments' or as gifts and were not part of daily consumption. Participants who emphasized the price rather than the Fair Trade quality explained their price-sensitive behaviour with doing shopping for a whole family. In contrast, other participants indicated that they were willing to pay an additional price for Fair Trade products.

By means of the IDM, the relevance of the different attributes was elicited quantitatively. This was operationalized via analysing the first and repeatedly accessed attributes. As other authors have shown, information that was accessed first and more frequently is assumed to be more important (Lürssen, 1989; Zander and Hamm, 2010). Even though the results of both analyses differed slightly, it became obvious that the Fair Trade label and the production method (organic or conventional) were the most important attributes. The ordering of product price and protection of children varied according to indicator (Table 8.4).

The labels play an important role for Fair Trade purchasing. In the FGD it was mentioned that labelling was important in order to verify that the Fair Trade attributes are controlled: 'The logo is very important ... because otherwise, everybody could write on it "we are nice to our employees".' The results of the IDM underline the importance of a Fair Trade label.

Nevertheless, information that goes beyond the Fair Trade aspect was important, too. The product that only carried the TransFair label and did not offer any further information was hardly ever purchased (3.1%), although it was the cheapest of all products. The product that did not carry an organic label was bought least often, irrespective of the price, and 82% of participants selected a product with the organic label.

The results for the importance of labelling are in line with Pelsmacker *et al.* (2005, 2006) who found that labels were an important communication instrument for Fair Trade information. On the one hand, labels are significant in order to recognize a Fair Trade product as such. On the other hand, participants trust the Fair Trade quality if the product is labelled accordingly. These findings are in contrast to those of Russeau (2015), who found that Belgian consumers did not perceive Fair Trade labels on chocolate as credible.

Table 8.4. Most important attributes.

Attribute	Number of first views (percentage of first mouse clicks)	Number of total views (percentage of all mouse clicks)
Fair Trade label	24.94	15.36
Production method (organic/conventional)	21.08	16.23
Protection of children	16.20	14.00
Product price	11.05	17.29
Social projects on-site	11.05	11.96
Trade relations	9.25	12.32
Producer income	6.43	12.84

The product that was purchased most often (44.7%) in both groups was coffee A, which included information about the Fair Trade aspects in an unspecific manner. The product furthermore carried the TransFair label and also the German organic label. With a price of €5.99 per 500 g [1] it was the second most expensive product. Even though the price was an important purchase criterion, it can thus be stated that it was not the cheapest product that was bought most frequently. The results are in line with those of Campbell *et al.* (2015), who found that that the higher price of Fair Trade coffee compared with conventional ones was perceived as being justified and did not have a negative impact on purchase intentions. The authors explained this phenomenon with consumers having a 'feeling of contributing to something good' and a 'warm glow' effect. The same effect may explain the fact that participants were also willing to pay a premium for the Fair Trade coffee that was additionally labelled as organic product. In contrast to those results, Andorfer and Liebe (2015) found that Fair Trade coffee sales rose when the price was reduced. They argued that the reduced price might have attracted new customers who bought Fair Trade coffee for the first time, not for ethical reasons but simply due to the price offer, whereas our study focused on participants who already bought Fair Trade coffee.

The results partly coincide with those of Pelsmacker *et al.* (2005), who also discovered that price and taste play an important role in the Fair Trade purchase decision. Van Loo *et al.* (2016) also found price to be amongst the most important product attributes when making a purchase decision for sustainably produced coffee. The relation between preferences for organic and social standards (Lehnert, 2009) is supported by the present findings. It can be stated that for the majority of the participants, Fair Trade posed an extra that consumers buy according to situational factors such as a mood or feeling that they have enough money.

8.3.4 Extent and strategies of information search

For marketers it is helpful to know the amount and particularly the strategies that consumers apply in information searches, in order to design the presentation of information accordingly. This holds particularly true for information provision on the internet, which is often designed in a matrix-like manner.

IDM experiments are especially helpful to give insights into the extent and strategies of information search. Due to the large amount of information available, consumers apply various strategies in order to reduce the cognitive burden associated with information processing. The IDM conducted in this research showed that consumers did not open all fields. The IDM contained in total 63 information fields that could be opened repeatedly. In this experiment, the extent of the participants' information search varied largely. On average, 51 fields were opened. However, a high standard deviation (26.70) implied that there was quite a spread between a number of participants who searched for information extensively and others who hardly searched for any information.

Theory suggests that certain strategies are applied in order to process information if people are confronted with bulky information (e.g. Payne *et al.*, 1978; Beatty and Smith, 1987; Ball, 1997; Sauermann, 2004). The analysis of search strategies was done with the help of a computer-based tool that visualizes the fields opened by the participants.

Search strategies can be grouped into two categories: compensatory versus non-compensatory strategies and alternative-wise versus attribute-wise strategies (Payne *et al.*, 1978, 1993; Ball, 1997; Sauermann, 2004). In the case of compensatory strategies, a trade-off between the product attributes is performed; for example, price is weighed against the organic quality. If non-compensatory strategies are applied, trade-offs are not made. Consequently, compensatory approaches are cognitively more demanding. Information search can also be performed either attribute-wise or alternative-wise. If an attribute-based search strategy is applied, the attributes of all product alternatives are regarded and compared in a sequence. If all attribute specifications of one specific product are regarded one after the other, an alternative-wise approach is applied. Additionally, mixed forms of attribute- and alternative-based information search can be observed. Attribute-wise approaches are considered cognitively less burdening (Russo and Dosher, 1983).

The results show that most of the participants (63%) applied attribute-wise strategies, which means that all products are checked first for one attribute and then for another and so on. About 20% of participants applied alternative-wise strategies, which implies that all attributes of one specific product are looked at first, followed by the next product with all its attributes. About 18% of the participants searched rather randomly, or had shifts in search patterns that could not be explained and did not seem to follow any rule.

It was also analysed whether participants used compensatory or non-compensatory strategies. Compensatory strategies require a higher cognitive effort, because trade-offs between attributes are made (Payne *et al.*, 1993). The results of the strategy analysis showed that compensatory and non-compensatory strategies were applied about equally often. Summing up, strategies with a lower cognitive effort (attribute-wise strategies and non-compensatory strategies) were strongly preferred. The identified search strategies were assigned to those commonly described in the relevant literature.

Together with the insights on the extents of the information search, we can conclude that even though time for purchase decisions is limited, Fair Trade consumers do indeed make an effort to gather pre-purchase information on Fair Trade products that are not yet known to them. At the same time consumers tend to limit their effort, especially as regards the cognitive involvement. It needs to be considered that the present study dealt with a product that was unknown to the consumers and in an experimental setting. The search effort for already known products but also for Fair Trade products under a well known brand will be significantly lower.

8.4 Conclusions and Recommendations for Marketers

The information given on product packaging helps initially and basically to inform consumers about the Fair Trade aspects of the relevant product. It can be assumed that usually it is not read again when purchasing the same product subsequently since it would be too time consuming. The results of the IDM experiments showed that consumers do search for information on the Fair Trade aspects that go beyond the mere label. Consequently, marketers should consider displaying additional information. Nonetheless, time for information uptake during shopping is restricted, as is the space on packages. Therefore, it is recommended to offer additional information using other media, e.g. via internet. Adding extra information that can be read after the purchase has a strong trust-building effect on future purchase decisions, especially if it is background information about the producers, since this creates a personal relationship, as the focus group discussions have shown. A convenient way for an easy uptake of additional information is the use of quick-response (QR) barcodes, which give easy access to the internet for consumers.

In any case, the 'right' information should be displayed, i.e. the information that consumers ask for. Regarding the design of the information, it became clear that consumers are usually not interested in bulky information; instead the texts should be short and clearly structured, even if shortness is at the expense of preciseness. Buyers do not want to spend much time in information search and processing during the act of purchasing. Organic production is an important additional attribute of Fair Trade products. As many consumers were not aware of the fact that not all Fair Trade products are organic, it is important for organic Fair Trade suppliers to communicate this additional value.

Displaying a Fair Trade label is highly relevant, as the results have shown. At the same time, marketers should be aware of the fact that almost 25% of participants purchased products that carried a fictional label. This implies that label communication should be improved. The establishment of a governmental Fair Trade label (best at EU level) on the social conditions of production and trade would offer consumers more security and trustworthiness in terms of production and ease the marketing of Fair Trade products. It would furthermore help consumers not to 'drown in the flood of labels', as mentioned in the FGD.

Regarding the content of the information, marketers need to be aware that Fair Trade is not the only product attribute that is relevant in the purchase decision. The combination of a Fair Trade label with the EU organic label should be pushed forward, due to the high importance of the organic attribute, combined with a positive willingness to pay. At the same time, the added value of Fair Trade products that also carry an organic label should be communicated more strongly, since not all consumers understand the difference. Marketing of Fair Trade products should consider both motivation bundles – hedonistic and altruistic.

When it comes to pricing, it is important to keep in mind that one of the more expensive products was purchased most frequently. This product was equipped with both the Fair Trade label and the organic label. This is an indicator for consumers' additional willingness to pay for Fair Trade and organic product features.

The analysis of extent and strategies of information search revealed that consumers aim to reduce the cognitive burden of information search by taking up only part of the information offered. Therefore, considering strategies in information search and presenting information in line with specific consumer needs is a likely success factor in the future, particularly for internet marketing. Consumers should be offered the possibility to search systematically for product attributes instead of having to compare whole products (Zander and Hamm, 2012).

According to the results of the case study research presented in this chapter, labels are the most important information source for consumers when looking for Fair Trade products. This is mainly due to consumers' desire to simplify decision making at the point of purchase. Consumers' preferences for labels were so pronounced – and knowledge so low – that they even bought products with fake labels. In addition to a label, concise information is appreciated, even if brevity is at the expense of preciseness. Organic production is an important additional attribute of Fair Trade products and many consumers are not even aware of the fact that not all Fair Trade products are also organic.

Price has been found to be relevant for the purchase decision for Fair Trade products, but participants were willing to trade off price for ethical attributes such as Fair Trade, organic and protection of child labour, particularly if these attributes were well communicated.

Note

(1) In worldshops the packet size was only 250 g due to the higher price levels. However, the absolute prices were the same.

References

Andorfer, V. and Liebe, U. (2015) Do information, price, or morals influence ethical consumption? A natural field experiment and customer survey on the purchase of Fair Trade coffee. *Social Science Research* 52, 330–350.

Aschemann-Witzel, J. and Hamm, U. (2011) Measuring consumers' information acquisition and decision behavior with the Computer-Based Information-Display-Matrix. *Methodology* 7, 1–10.

Bailey, K. (1994) *Methods of Social Research*. Free Press, New York.

Ball, C. (1997) A comparison of single-step and multiple-step transition analyses of multiattribute decision strategies. *Organizational Behavior and Human Decision Processes* 69, 195–204.

Beatty, S.E. and Smith, S.M. (1987) External search effort: an investigation across several product categories. *Journal of Consumer Research* 14, 83–95.

Beccheti, L. and Rosati, F. (2005) The demand for socially responsible products: empirical evidence from a pilot study on fair trade consumers. *Society for the Study of Economic Inequality* 4, 1–40.

Bryman, A. (2008) *Social Research Methods*. Oxford University Press, Oxford, UK.

Campbell, C., Heinrich, D. and Schönmüller, V. (2015) Consumers' reaction to fair trade motivated price increases. *Journal of Retailing and Consumer Services* 24, 79–84.

Daniel, M., Bricas, N. and Sirieix, L. (2010) L'intérêt pour les petits producteurs fonde-t-il consommation des produits equitables? *Cahiers Agricultures* 19, 28–33.

Ferran, F. de and Grunert, K. (2007) French fair trade coffee buyers: purchasing motives: an exploratory study using means-end chains analysis. *Food Quality and Preference* 18, 218–229.

Gläser, J. and Laudel, G. (2006) *Experteninterviews und Qualitative Inhaltsanalyse*. Verlag für Sozialwissenschaften, Wiesbaden, Germany.
Halkier, B. (2010) Focus groups as social enactments: integrating interaction and content in the analysis of focus groups data. *Qualitative Research* 10, 71–89.
Hwang, M.I. and Lin, J.W. (1999) Information dimension, information overload and decision quality. *Journal of Information Science* 25, 213–218.
Janssen, M. and Hamm, U. (2012) The mandatory EU logo for organic food: consumer perceptions. *British Food Journal* 114, 335–352.
Lamnek, S. (2005) *Gruppendiskussion – Theorie und Praxis*. Beltz, Weinheim, Germany.
Lehnert, M. (2009) *Präferenzanalyse Ethischer Produkte: Eine Verhaltenswissenschaftliche Analyse am Beispiel von Bio und Fair Trade*. Dr. Kovac, Hamburg, Germany.
Lürssen, J. (1989) *Produktwissen und Kaufentscheidung: Einbeziehung des Produktwissens bei der Analyse von Kaufentscheidungen mit der Informations-Display-Matrix*. Lang, Frankfurt, Germany.
Mayring, P. (2008) *Qualitative Inhaltsanalyse: Grundlagen und Techniken*. Beltz, Weinheim, Germany.
Padel, S. and Foster, C. (2005) Exploring the gap between attitudes and behaviour. Understanding why consumers buy or do not buy organic food. *British Food Journal* 107, 606–625.
Payne, J.W., Braunstein, M.L. and Carroll, J.S. (1978) Exploring predecisional behavior: an alternative approach to decision research. *Organizational Behavior and Human Performance* 22, 17–44.
Payne, J.W., Bettman, J.R. and Johnson, E.J. (1993) *The Adaptive Decision Maker*. Cambridge University Press, Cambridge, UK.
Pelsmacker, P. de, Janssens, W., Streckx, E. and Mielants, C. (2005) Consumer preferences for the marketing of ethically labelled coffee. *International Marketing Review* 22, 512–530.
Pelsmacker, P. de, Janssens, W., Sterckx, E. and Mielants, C. (2006) Fair-Trade beliefs, attitudes and buying behaviour of Belgian consumers. *International Journal Non-Profit and Voluntary Sector Marketing* 1, 125–138.
Russeau, S. (2015) The role of organic and fair trade labels when choosing chocolate. *Food Quality and Preference* 44, 92–100.
Russo, J. and Dosher, A. (1983) Strategies for multiattribute binary choice. *Journal of Experimental Psychology: Learning, Memory, and Cognition* 9, 676–696.
Sauermann, H. (2004) Vocational choice: a decision making perspective. *Journal of Vocational Behaviour* 66, 273–303.
Schleenbecker, R. and Hamm, U. (2015) Information needs for a purchase of fair trade coffee. *Sustainability* 7, 5944–5962.
Tagbata, D. and Sirieix, L. (2008) Measuring consumer's willingness to pay for organic and fair trade products. *International Journal of Consumers Studies* 32, 479–490.
Van Loo, E., Caputo, V., Nayga, R., Seo, H., Zhang, B. and Verbeke, W. (2016) Sustainability labels on coffee: consumer preferences, willingness-to-pay and visual attention to attributes. *Ecological Economics* 118, 215–225.
Verbeke, W. (2005) Agriculture and the food industry in the information age. *European Review of Agricultural Economics* 32, 347–368.
Wright, L. and Heaton, S. (2006) Fair trade marketing: an exploration through qualitative research. *Journal of Strategic Marketing* 14, 411–426.
Zander, K. and Hamm, U. (2010) Consumer preferences for additional ethical attributes of organic food. *Food Quality and Preference* 21, 495–503.
Zander, K. and Hamm, U. (2012) Information search behaviour and its determinants: the case of ethical attributes of organic food. *International Journal of Consumer Studies* 36, 307–316.

9 Gender-equality Chocolate: a Missing Market?

Casey Goldvale[1]* and Romane Viennet[2]
[1]*Georgetown University, Washington, DC, USA;* [2]*Cornell University, Ithaca, New York and Sciences Po Paris, France*

9.1 Introduction

Since the institutionalization of the Fair Trade movement, a body of literature has grown to evaluate the certification programme's success. Recently, the literature has become so well explored that academics and non-governmental organizations (NGOs) alike have begun delving deeper into the supply side of value chains, finding which specific tenets of the Fair Trade social justice principles are being met and which are left unaddressed (Greig, 2006; Boersma, 2009; Bacon, 2010; Valkila *et al.*, 2010). However, from meta-analysis of the existing literature, researchers have not handled ethical consumption with the level of nuanced critique applied to Fair Trade production (Stolle *et al.*, 2005; Andorfer and Liebe, 2012).

Among the numerous trade-born issues that the international Fairtrade Labelling Organizations (FLO) aims to tackle, gender inequality in the supply chain is particularly pervasive and creates some of the most serious structural setbacks to economic development (Buvinic *et al.*, 2009). Despite this commitment, the Fair Trade system's benefits fail to reach a number of marginalized groups, including disadvantaged women (Greig, 2006; Bacon, 2010). Gender disparities, especially in the form of income inequality and resource access, are rampant in agriculture across the world (Quisumbing *et al.*, 2014). This is especially detrimental since it is well documented that female empowerment could benefit industries in developing nations, such as cocoa production, by making them more productive and profitable to smallholder producers (Laven, 2012).

In recent years, NGOs, academic researchers, private sector markets and policy makers have been discovering and acting upon a growing body of evidence that women are a drastically underutilized key to economic development. Giving women leading roles improves success and profitability across a variety of fields (Hickey, 2014; Noland *et al.*, 2016). Particularly within international agriculture, closing the gender gap in yields would reduce the number of undernourished people by 100–150 million and could increase agricultural output in developing countries by between 2.5% and 4% (FAO, 2011).

Investing in women also produces multigenerational effects. Upon receiving aid, loans, or profits, women in developing countries tend to spend the funds on their household and community needs, whereas men are more likely to engage in personal spending on their own behalf. Additionally, studies performed in countries across the globe show that when women have more agency and resources within a household

*Corresponding author; e-mail: cmg275@cornell.edu

or community, rates of poverty and child malnutrition drop while rates of women's self-employment and child education rise (IIPS, 2007; Menon *et al.*, 2016). In a more typical household with a male head, daughters especially lag behind in support and development. Thus, investing in a single generation of women is also an investment in their daughters, which has far-reaching household and community level intertemporal effects on hunger, poverty, education and inequality. Key policy makers have even gone so far as to declare interventions empowering women are 'silver bullet[s]' (UN Women, 2015).

Despite the increasing recognition of the developmental importance of gender equality, women in international agriculture face a slew of obstacles to success. Though women are integral to growing and harvesting efforts, making up 66% of agricultural work, producing 50% of crops, they manage only 10% of income and own merely 1% of land. In their work, they make far less than male counterparts doing similar labor (Laven, 2012). Poverty among rural women is of great concern where over the past '21 years, in 41 developing countries poverty among rural women has risen by 17 per cent' (Hutchens, 2010). Even those in management face difficulties; a 2011 report from the FAO identified a significant gender disparity regarding access to capital inputs, such as credit, land and machinery, which has led to a 10–20% reduction in crop yields for women compared with men. Even Fair Trade cooperative organizations, which are typically intended to empower those marginalized by supply and value chains, lack female participation (Lyon, 2008).

Historically, certifications have done little to address the issue. Recent research from Fairtrade International found that 'women's membership of Fair Trade Small Producer Organisations (SPOs) around the world is just 22 per cent' (Fairtrade International, 2015). According to Hutchens (2010), FLO's Strategic Funding Review reveals that only 10% of the board members for producer cooperatives and Joint Bodies for plantation workers are women. 'These gendered roles affect women's ability to benefit from Fair Trade equally, not only in terms of representation but also of the additional benefits that accrue to co-operative leaders, such as market contracts and information, skills and decision-making influence' (Hutchens, 2010).

High valuation of gender equity is not just limited to a subset of international development. A survey run by the Pew Research Center in 2015 found that gender equality is the second most valued civil liberty to individuals across the world, second only to religious freedom. On average, 65% of global respondents ranked gender equality as highly important; in the USA and Canada, the same was found true of over 90% of respondents. The development benefits implied by empowered women, their role in agriculture across countries and this high demand for gender equality are among many reasons to investigate whether certification can support female empowerment in the way that it does the environment and Fair Trade.

Certification efforts are starting to respond to national demand. A new Swiss-based private label called Economic Dividends for Gender Equality (EDGE) seeks to highlight companies with best practices regarding gender treatment. Fairtrade International has also adopted a Gender at Work framework that incorporates gender into the global strategy of the organization. Though these efforts are too recent to have been subjected to extensive academic review, research shows that identifying and empathizing with producers is typically a more powerful motivation to Fair Trade consumers than simple altruism (Doran, 2010); thus, it would be fair to hypothesize higher success for a certification guaranteeing gender equality than for Fair Trade alone, at least among women and those who value gender equality highly. This chapter takes the conversation a step further and aims to investigate how consumers will respond to a gender equality label. It was written as a case study based on Basu and Hicks (2008).

We conducted a market analysis to determine the premium that consumers may be willing to pay for a label addressing supply chain gender equality in its own right. We observed typical Fair Trade products in local grocery stores and cafés and chose to focus on chocolate production through our gender equity lens. Chocolate is a logical choice, because there is a strong premium chocolate market whose demand levels have proven relatively robust to changes in consumer behaviour due to economic downturns (Haddon and Rai, 2012). Also, women play critical roles in two-thirds of the production stages that determine cocoa quality and so demand for female labour is likely to be more inelastic to increased labour value than may be the case with other crops. Moreover, 90% of cocoa is produced by smallholder farmers; this structure provides more opportunities for female management than a larger scale and lends itself better to cooperatives for Fair Trade style certification (UTZ, 2009).

We surveyed 71 students on the Cornell University campus in Ithaca, New York, and used the stated preference method to elicit their willingness to pay. More specifically, we disentangled the effect of different types of information – such as premium range or country of origin – on respondents' willingness to pay. This gave us some indications as to which composition optimizes the price premium yielded by the product. Our methodology is similar to that of Basu and Hicks (2008) but, rather than viewing the existing market for Fair Trade coffee, we shifted our perspective to a potential market for chocolate bearing a certification of gender equality in cocoa production. We found that: (i) there exists an unexploited niche market for gender-equality labelled chocolate(1) among the student community with a price premium of at most $1 per 3.5 oz (85 g) bar; and (ii) information about the country of origin of the cocoa, the certifying organization and the impact of the gender-equality programme behind the label all have a strong, positive effect on labelled product take-up. Additionally, though we cannot make causal assertions, our descriptive statistics suggest that the most likely consumer may be a woman in an affluent, politically conscious and active environment.

9.2 Methodology

We used the stated preference method (conjoint analysis) to engineer our survey and identify a potential market for our chocolate bar. Conjoint analysis is an advanced market research technique used to unveil the underlying preferences of consumers when they choose one product over a potential substitute (Green and Srinivasanm, 1990). Central to the survey, our choice experiment comprised four questions, which aimed to: (i) unveil which label composition maximizes probability of purchase; and (ii) elicit the optimal price premium at which a gender-equality chocolate bar could be sold. The four key questions of our survey implored subjects to choose one chocolate bar in a panel with three 'buy' options and one 'opt out' option, where each three-option set varied different attributes.

Though it has been used widely in marketing research, conjoint analysis also has its limitations. Firstly, there is a trade-off between too many and too few attributes to display on the purchasing options. We chose to limit our analysis to six attributes and explain our selection in the following section. What is more, the analyses we carried out before selecting those attributes and the corresponding range of values should limit the risks of falling out of the ranges acceptable to our respondents. On the other hand, the literature often emphasizes the tendency to use 'simplifying heuristics' when responding to a conjoint analysis-type survey, for the decisions might be too complex to captivate the interest or intelligence of the respondents (Cunningham *et al.*, 2010). We tried to make the decision as simple as it could get, explaining the functioning of our hypothetical gender equality programme and limiting the number of attributes and the type of purchase required. Finally, by carefully selecting the attributes, we also reduced the risk of seeing respondents discard options based on a most-disliked attribute such as company brand or country of origin, and vice versa – both the attributes and the values we selected seemed close enough to those used in the actual chocolate markets (see next section).

9.3 Data Collection

We surveyed 71 Cornell graduate and undergraduate students. Our sample was split into two groups: 30 of the students were randomly picked among a pool of 60 members of intentional living communities(2); and 41 in one of the university's library cafés. The largest consumer age group of chocolate in the USA is the 18–25-year-old cohort, which is well represented by the average ages of the university student population, which is approximately 75% undergraduate and 25% graduate students (Cornell University, 2016). Additionally, through Cornell's admissions process, the campus is intentionally diverse, representing almost all US states and territories and with racial and ethnic makeup similar to the entire country. A higher percentage of Cornell undergraduates are Asian and a lower percentage Black than are counted in the US Census. Over 50% of undergraduate students receive need-based financial aid, which takes into account the cost of eating on Cornell's campus. Fewer than 10% of students are international. Though probably not perfect, randomly surveying consumers at the most popular café on such a campus yields a sample diverse enough to get a sense of some potential market characteristics for gender-equal chocolate.

As seen in Table 9.1, breaking down the sample into the first 30 students surveyed, who lived in intentional communities, and the population selected from a popular campus café revealed differences between the two groups' propensities toward ethical consumerism and social justice causes. On average, the first 30 students placed higher value on the importance of gender issues and were more likely to be involved in activities surrounding gender issues. They also self-reported greater frequency of purchasing certified foods. These characteristics are supported by research findings that show individuals who live in communities with higher social capital are more likely to be political consumers (Neilson and Paxton, 2010). Based on our findings and research on social capital, we therefore hypothesize that the first 30 students surveyed, who live in intentional communities, will be better informed about and have a higher willingness to pay for ethical consumption than the average consumer. In general, we assumed these students were also slightly more social-justice conscious than the average student, though they may not necessarily identify as feminist or be specifically involved with women's rights.

The survey started with brief information about the cocoa supply chain, an overview of the situation of gender equality in the cocoa production industry of each of the three countries of origin, and a short explanation of the gender-equality programme our label suggested. The twelve questions were distributed as follows: eight questions asked general information on the respondent characteristics (gender, education, involvement with gender inequality issues, consumption habits, general attitude toward labelled products etc.) and four investigated consumer preferences.

We fixed the weight of the bar to 3.5 oz (85 g) and also fixed the proportion of cocoa to 70% for each bar, since a benchmark survey we carried out on the brands sold on the campus and in the town where we performed our analysis showed that certified chocolate bars had on average a 70% cocoa proportion advertised on their packaging and 3.5 oz (85 g) is the most common size for a chocolate bar. The same benchmark showed that a large share of the chocolate offered on the local market was labelled 'organic,' especially among chocolates with different labels, such as Fair Trade.

To remain consistent with the consumer landscape we were facing, we decided to reflect this pre-eminence on our choice experiment – hence the presence of the USDA organic label on

Table 9.1. Description of the sample (with average data in columns 2 to 7).

	No. of obsv.	Importance of gender amongst other issues[a]	Level of involvement with gender issues[b]	Belief in effectiveness of ethical consumerism	Frequency of purchase of certified food[c]	Frequency of purchase of dark chocolate[c]	Part of budget allocated to chocolate[d]
Overall sample	**63**	**2.57**	**3.35**	**2.10**	**1.92**	**1.61**	**1.87**
Female	35	2.39	3.34	2.26	2.00	1.83	2.14
Male	24	2.76	3.36	1.88	1.71	1.29	1.54
Undisclosed	4						
By group							
Intentional living communities[e]	26	2.81	3.19	2.32	2.11	1.70	1.78
Café patrons[f]	37	2.43	3.46	1.97	1.80	1.54	1.94

[a]Scale of 1–5 (1 = most important). [b]Scale of 1–4 (1 = most involved). [c]Scale of 0–3 (3 = often). [d]Scale of 1–5 (5 = entire budget). [e]Surveyed 19 April 2014. [f]Surveyed 25 April 2014.

all three options in each panel. This also allowed for comparing very close substitutes and thus for more accuracy in assessing the impact of the single gender-equality label. As for the varying attributes, we selected the ones that we observed most commonly on the packaging of labelled chocolate bars, i.e. country of origin of the cocoa, certification logo, price, information about the programme's impact when applicable and certifying agency. In addition to our observations, the qualitative study performed in North America by Boivin *et al.* (2012) found that country of origin and certification logo counted among the most consequential label attributes. We decided to hold constant the other two attributes that they studied: chocolate bar's cocoa content (70%) and weight (3.5 oz (85 g)). Information about the programme's impact aimed to meet consumers' apparent preference for label transparency about producer benefits (Basu and Hicks, 2008). Table 9.2 gives a detailed description of each attribute and a justification for the options we selected. The price determines the premium we are trying to estimate in order to determine the value of our product on the chocolate market. The range is voluntarily loose, and large in magnitude to fit the reality of the labelled-product market.

The relationship between the premium and the impact on female grower revenue is structured as follows: there is always a minimum portion withheld on the premium ($0.50 in panels 4, 5, 7 and $0.25 in panel 6) to account for various costs of implementing the programme, while the remaining portion ($0.75 in panels 4, 6, 7 and $1.00 in panel 5) is assumed to go direct to female producer revenue.

We approached the students with a carefully respected scenario. If the individual agreed to take the survey, we handed it to them with a pen. If the individuals were in a group, we would always address each of them individually, and state that they did not have to take the survey even if the other members of the group did. No other information was given orally, except the definition of one term, 'ethical consumerism', given to one respondent as follows: 'ethical consumerism is a form of social activism that uses the consumer's purchasing power as a means to support a cause'. The respondents took on average 6 min to complete the survey.

9.4 Empirical Strategy

Our analysis revolves around a conditional logistic estimation of the likelihood of buying chocolate, regressed on the product attributes and grouped by respondent. We use this model to compute the actual probability of purchase. With respect to our hypothesis regarding our sample's composition, we also run a separate conditional logit (clogit) estimation for each group, differentiated by date. This discrimination will allow us to see whether there actually exists a stronger market for gender-equality chocolate among intentional living communities. The four-by-three choice experiment generated 12 observations for each participant. Since individual characteristics do not vary between entries for each respondent, we used two ways to control for the most relevant individual characteristics: (i) a logit estimation with fixed effect controlling for gender, perception of ethical consumerism efficiency, involvement with gender equality, buying habits; and (ii) descriptive analysis to display the frequency of purchase for each above-mentioned control.

Our clogit estimation on a consumer choice, modelling willingness to pay based on price and attribute impact, is as follows and is grouped by respondent ID:

$$\begin{aligned} choice = {} & \alpha 0 + \alpha 1 price + \alpha 2 coo\_ci \\ & + \alpha 3 coo\_indo + \alpha 4 impact \\ & + \alpha 5 icco + \alpha 6 unww + \varepsilon \end{aligned} \quad (9.1)$$

where *choice* represents the respondent's discrete choice between chocolate A, B or C for each of the four panels, *coo_ci* and *coo_indo* are the cacao countries of origin, *impact* is the impact of the gender equality label, i.e. the increase in female cacao grower revenue due to the purchase of labelled chocolate, and *icco* and *unww* are two of the alternatives for certifying agencies.

We then selected four individual characteristics that may shed light on the profile of the most responsive consumer. Our analysis of these characteristics relied on observations of distributions and descriptive statistics; therefore, we did not make causal claims, but rather identified whether correlations existed that were worth future investigation. We hypothesized that respondents' gender (*female*=1) was of interest since we expected women to be more likely to value gender equality strongly (Pew Research

Table 9.2. Description of the attributes.

Attribute	Range of values	Value for the non-labelled baseline	Justification
Price	$2.50, 3.50, 3.75, 4.00	$2.50	The prices were based on a price study we did on the labelled chocolate bar market accessible to our population (we used price catalogues from grocery stores). A 3.5 oz (85 g) chocolate bar with two labels (organic and Fair Trade, organic and Rainforest etc.) was sold between $3.50 and $3.75 on average.
Gender equality label	Present/absent	Absent	The gender equality label was a very visual way to attract respondent's attention to the fact that the chocolate she was choosing supported gender equality or not.
Impact of gender equality programme on female grower revenue	$0.50, 0.75, 1.00 (when applies)	N/A	Impact on female grower revenue was related to both the label and the price premium. It gave more information on the actual effect of the gender equality program. Our assumption was that the respondent's willingness to pay (i) may be revealed more clearly if she had information about the actual impact and (ii) may vary depending on how much of the price premium actually went to the female producer. Our estimations of the basic female revenue ($0.50/lb) represents around 10% of the international price/lb.
Country of origin (cacao)	Côte d'Ivoire, Indonesia, Honduras	Honduras	We picked one country among the world's top producers, from each of the regions of the world that are involved in cocoa production. By doing so we wanted to be able to control for the respondent's potential preference for one country/region over the other. Furthermore, we picked cocoa-producing countries where studies have shown that gender inequality is particularly salient in rural areas, specifically among cocoa producers.
Certifying agency	International Cocoa Organization (ICCO), United Nations Women Watch (UNWW)	N/A	We picked two certifying agencies that we clearly identified in the survey's introductory paragraph as being more or less concerned with gender inequalities. ICCO would reveal (and control for) a preference for cocoa quality over supporting gender equality while UNWW would reveal the opposite preference.
Organic label	Organic	Organic	Since most of the labelled chocolate we found on the market was double labelled 'organic' and something else (Fair Trade, Rainforest, vegan …), and since we wanted our panels to propose very close choice options, we decided to hold the organic attribute fixed across all choices.

Center, 2015). If this were the case, it would likely follow that being a woman increased the probability to take up a price premium. We also assumed that the difference in the price premium between male and female, given that the respondent took up a premium, was positive (i.e. women would more often take up a larger price premium than men). The motivation in putting out a gender-equality chocolate bar on the market was to further capitalize on a market niche composed of ethical consumers. However, we were interested in knowing to what degree ethical consumerism correlated with product uptake, and whether the product would attract consumers who were not necessarily firm believers in ethical consumerism (*eff_consume*=1). The same rationale motivated our choice of controlling for the respondent's involvement with promoting gender equality. If the respondent was already very involved (*involve*=4), we predicted that they were more likely to take up our product with a higher premium. Our assumption corresponded to previous literature finding a positive effect of political involvement on political consumption (Andersen and Tobiasen, 2004; Forno and Ceccarini, 2006; Shah *et al.*, 2007). Finally, we investigated the respondent's consumption habits regarding dark chocolate to gain some intuition as to whether an already defined affinity for dark chocolate had a relationship with willingness to pay and product uptake. We presented the distribution of choice by gender, perception of efficiency, involvement and consumption habits in four different graphs.

Table 9.3. Likelihood of choosing one chocolate bar over the other two options.

	Choice (1)[a]	Choice (2)[b]	Choice (3)[c]
Price	–4.643	–3.522	–5.481
	(1.469)**	(2.214)	(1.970)**
Country of origin	1.174	1.233	1.139
Côte d'Ivoire	(0.300)**	(0.449)**	(0.409)**
Country of origin	0.841	1.115	0.622
Indonesia	(0.324)**	(0.478)*	(0.454)
Impact	2.064	–0.743	4.121
	(1.657)	(2.551)	(2.203)
Certifying agency	3.047	3.807	2.509
ICCO	(0.866)**	(1.328)**	(1.155)*
Certifying agency	4.386	5.553	3.567
UNWW	(0.824)**	(1.303)**	(1.084)**
N	717	303	414

[a]Results of main conditional logistic regression; [b]results of conditional logistics regression for respondents from intentional living community; [c]results for respondents outside the intentional living community; $*P < 0.05$; $**P < 0.01$.

9.5 Results

9.5.1 Determinants of the probability of purchase

The results of the estimation are summarized in column (1) of Table 9.3. The results of the estimation for each subsample identified in the sample methodology can be found in columns (2) (intentional living community members) and (3) (café students).

The coefficients in Table 9.3, column (1), give the contribution of each product attribute to the likelihood that the respondent chooses one of the three chocolates – be it labelled or not. These coefficients give a sense of the impact of each attribute on the decision-making.

Price has a large, negative impact on the probability to buy any of the chocolate bars (–4.643). The programme impact has a positive effect on the probability of purchase, although smaller in magnitude than the price effect. This seems to imply that consumers value more greatly their own purchasing power (which is reduced when price increases) than the impact that the programme has in terms of revenue increase for female producers. Respondents seem to trust the United Nations Women Watch (UNWW) certification over that of the International Cocoa Organization (ICCO). Côte d'Ivoire as the country of origin seems to have the greatest, positive impact on the probability of purchase compared with other countries of origin. This bodes well for the programme's potential for positive impact on the production side, since, of the three possible countries of origin, Côte d'Ivoire is the greatest gender inequality offender according to UNDP's 2013 Human Development Report.

Our hypothesis of a strong niche existing among more social justice-conscious consumers was mostly confirmed. Comparing the results in columns (2) and (3) of Table 9.3 shows that the negative effect of price reduces in magnitude and significance for the intentional living community group. This suggests that consumers who are more social-justice conscious and already participate in alternative consumption practices are less price sensitive when it comes to supporting gender equality. On the other hand, their probability of buying a chocolate bar is negatively affected by the impact reported for the programme. This result is unexpected and less significant than the other attributes included in our model. An optimistic explanation is that, since very few other ethical consumerism initiatives focus on gender equality, members of cooperative living communities are concerned about participating in the initiative, irrespective of this initiative's impact.

9.5.2 Correlations between choice and consumer characteristics

We base our investigation of chosen profile characteristics on observing descriptive statistics and basic correlations.

Options in the first panel (as shown in Fig. 9.1) are denoted by 4a, 4b and 4c; then 5a, 5b and 5c in the second panel; 6a, 6b and 6c in the third; and 7a, 7b and 7c in the fourth. For each panel, choices 'a' and 'b' include premiums for the label.

Figure 9.2 shows the number of respondents who chose each panel option, by gender. For panels 5, 6, and 7, male and female responses followed similar patterns. In panel 5, 'b' was the most popular choice, followed by 'c'. In panel 6, 'a' was the most popular, followed by 'c'. In panel 7, 'c' was the most popular, with 'a' and 'b' tied for second place. However, in panel 5, women exhibited a much stronger preference for the chocolate with the gender equality label than men. Additionally, although each gender generally ranked the options similarly, women made up a significantly higher proportion (62% on average) of the total that would pay a premium for the gender equality label than men, even taking into account that women made up 59% of the sample.

As Fig. 9.3 displays, respondents who chose option 'b' had systematically a higher faith in the effectiveness of ethical consumerism. For each panel, option 'b' was the labelled chocolate bar with the lower premium. These results suggest that faith in ethical consumerism encourages consumers to pay extra money for the gender equality certification, but not necessarily to choose the highest premium option. However, in

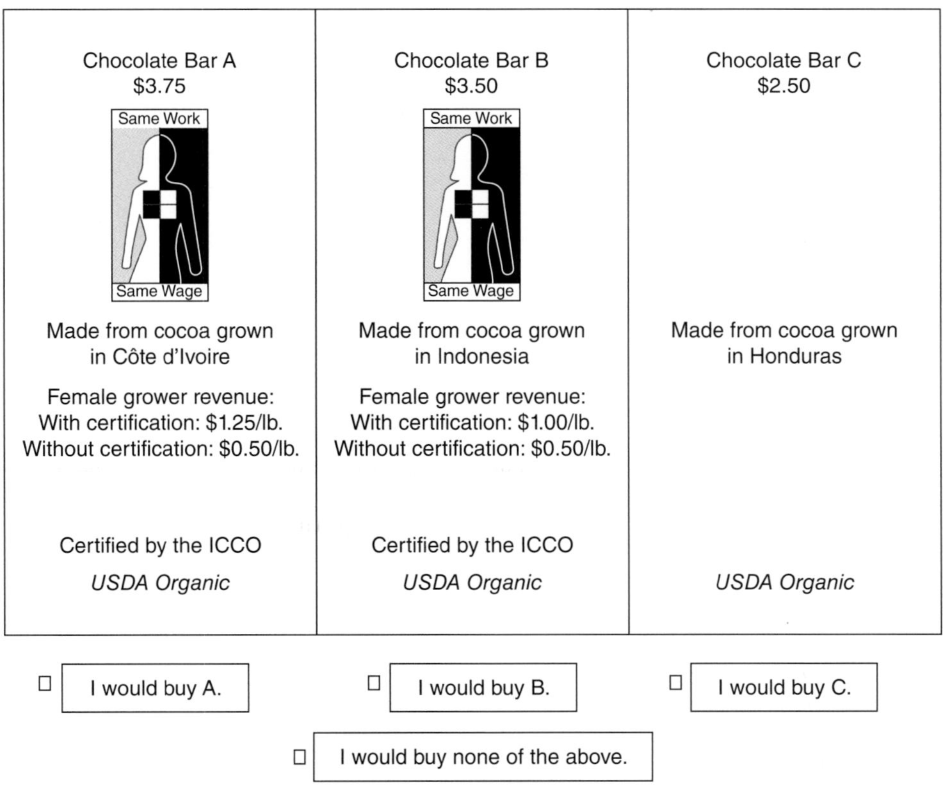

Fig. 9.1. Example of panel 1 from the survey.

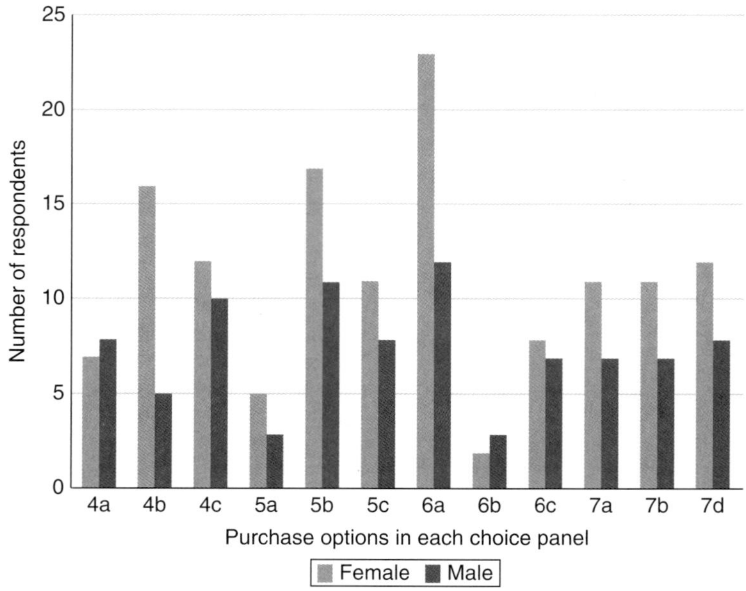

Fig. 9.2. Number of respondents per option in each panel, by gender.

the third panel, option 'a' was by far the most popular choice, even though Fig. 9.2 demonstrates that those participants averaged a lower faith in the efficiency of ethical consumerism. For this panel, two labelled options ('a' and 'b') were given with the same premium, but differed in the amount of female grower revenue listed on the label.

Figure 9.4 displays a positive correlation between willingness to pay and level of involvement with gender equality issues. In each panel, the participants who chose the bar with the highest premium had the highest average involvement. The participants who chose the bar with the lower premium had the next highest average involvement. Those who chose the cheapest chocolate bar, which was unlabelled, demonstrated the least involvement.

Figure 9.5 demonstrates that people who already buy dark chocolate most frequently may be more willing to pay a premium for the labelled chocolate bars. In the first two panels, the participants who chose the option with the highest premium for gender equality certification reported the highest average frequency of buying dark chocolate. In the second two panels, the participants who chose the option with the lower premium amount reported the highest average frequency of buying dark chocolate.

9.5.3 Maximum acceptable premium estimation

If the 'gender equality in cocoa production' programme is to be implemented, we need to know the maximum premium our target customers are willing to pay when putting the product on the market. In order to do so, we calculate the probability of choosing the labelled chocolate bar with the higher premiums (chocolate A) in all panels based on an adapted type of fitted values obtained with the conditional logistic regression model (we do not take into account the constant). Inserting the coefficients estimated in Table 9.3 into Eqn 9.1 yields the following:

$$\hat{choice} = -4.643 * price + 1.174 * coo\_ci + 0.841 * coo\_indo + 2.064 * impact + 3.047 * icco + 4.386 * unww \quad (9.2)$$

where for the fitted values of choice in each panel ($\hat{choice}_{K,J}$, where $K = (A,C)$ and $J = (4,5,6,7)$), the regressors take the values corresponding to the alternative option (C or A). Then, using the conditional logit expansion of the multinomial logit model, the

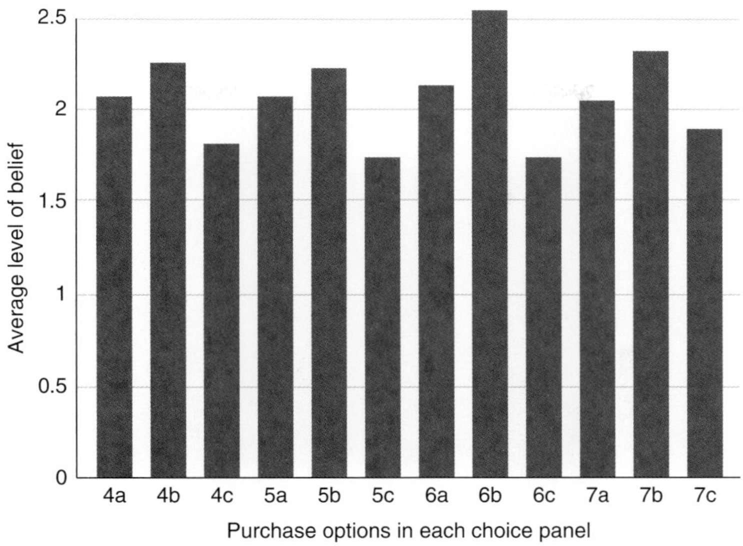

Fig. 9.3. Respondents' average level of belief in effectiveness of ethical consumerism per option in each panel.

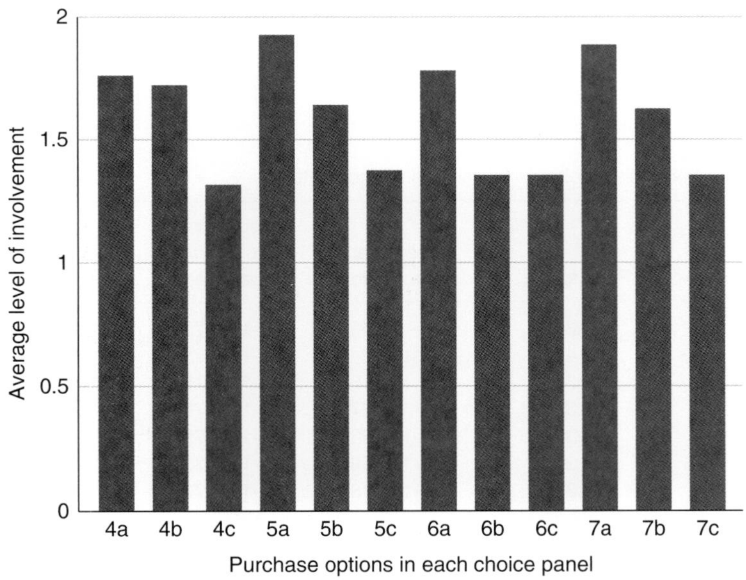

Fig. 9.4. Respondents' average level of involvement with gender equality issues per option in each panel.

probability of choosing chocolate A in every panel is:

$$p_A = \left(\frac{EXP\left(ch\hat{o}ice_{A,J}\right)}{\sum EXP\left(ch\hat{o}ice_{K,J}\right)} \right) \tag{9.3}$$

This estimation yields the results reported in Table 9.4. The probability of choosing 'A' – the option that always yields the highest posited premium – peaks at 0.78 for the estimation in panel 6. This means that the combination of attributes for chocolate 'A' in panel 6 is the optimum if we want to maximize the take-up of this chocolate. Unsurprisingly, this optimum is correlated with the smallest premium of $1, which means that anyone who wants to put a

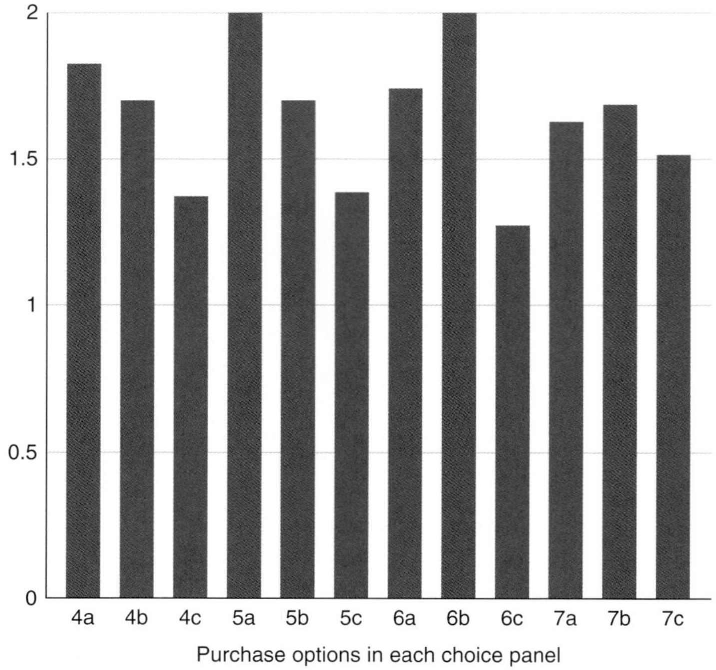

Fig. 9.5. Respondents' average frequency of purchasing dark chocolate per option in each panel.

Table 9.4. Estimation of the probability of choosing chocolate A.

Choice panel	Price for A ($/lb)	Prediction of *choice* for A	Prediction of *choice* for C (baseline)	Premium ($)	Probability of purchasing A
Panel 4	3.75	−11.6423	−11.6075	1.25	0.4913
Panel 5	4	−12.62	−11.6075	1.5	0.2665
Panel 6	3.5	−10.3165	−11.6075	1	0.7843
Panel 7	3.75	−11.4773	−11.6075	1.25	0.5325
Other premiums	3.51	−10.3629	−11.6075	1.01	0.7764
based on Panel 6	3.52	−10.4094	−11.6075	1.02	0.7682
characteristics	3.55	−10.5487	−11.6075	1.03	0.7425
Characteristics of chocolate A in Panel 6					
COO	Honduras				
Impact	0.75				
Certifying agency	UNWW				

3.5 oz (85 g) gender-equality chocolate bar on the market should aim for a premium of $1. Checking for several price levels between $3.5 and the price yielding the second highest premium ($3.75), we can confirm that any premium above the $1 threshold yields a lower probability of purchasing.

9.6 Summary, Conclusion and Policy Implications

Our analysis of consumer stated preferences and self-reported individual characteristics demonstrates that there exists a niche market for gender-equality chocolate bars. In addition, the experiment reveals details about how to optimize the label design and provides insight into potential consumer populations worthy of more in-depth market research. Such a label programme could provide innumerable benefits, socially and economically, by promoting proper recognition and compensation for the contributions of women as decision makers and workers crucial to global agricultural production.

We observed that some attributes had more impact on consumer willingness to pay than others. Firstly, the current analysis shows that, for dark chocolate with a gender-equality label, price increases negatively affect interest in the chocolate product. The existing literature on ethical consumerism shows mixed findings on sensitivity to higher prices. For instance, Hainmueller *et al.* (2014) found that most consumers were not sensitive to higher prices for Fair Trade labelled coffee, but there was a segment of consumers that would not pay a large premium for labelled products. On the other hand, Andorfer and Liebe (2015) and Basu and Hicks (2008) both found evidence of consumer price sensitivity for the labelled goods that they studied. Despite this negative price effect, participants were nevertheless willing to pay the premium for products with additional attributes. Of all the price premiums associated with the hypothetical label presented to survey respondents, chocolate bars with approximately a $1 premium had the highest probability of purchase, exceeding 75%. About 50% of consumers were still willing to pay a premium of $1.25, but probability of purchase dropped to below 30% with a premium of $1.50. Our descriptive investigation into correlations therefore suggests that it is possible to enable the product to reach a wider audience and nudge consumers who demonstrate low faith in ethical consumerism to pay a premium for a labelled chocolate bar of $1, with the right combination of attributes explained on the label.

The information about the programme impact on female producers has an effect that is approximately half the price effect. It follows that producers wishing to introduce a gender-equality centred label could mitigate consumer price sensitivity by listing the increase in female grower revenue. This effect was nevertheless negative among members of intentional living communities: an optimistic explanation is that, since very few other ethical consumerism initiatives focus on gender equality, these respondents were concerned about participating in the initiative, irrespective of this initiative's impact. It would appear that, at least to launch the programme, participating producers should be from Côte d'Ivoire, since participants were especially responsive to products made from cocoa grown in that country. However, we highly doubt that all respondents knew the exact rank of Côte d'Ivoire relative to Honduras and Indonesia in terms of gender inequality, so we believe this reveals a strong preference either for the country or for the quality of the cocoa from Côte d'Ivoire itself. This preference indicates it would be beneficial if the programme were implemented in Côte d'Ivoire, since more consumers would buy Ivorian cocoa-based chocolate. A worthwhile endeavour for any adoption of a gender-equality certification would be to further identify locations of production where consumer preference and economic marginalization of women coincide. Programmes should nevertheless avoid marginalizing even more female producers from the less popular countries of origin, for instance by working with them on enhancing their cocoa quality, access to beneficial supply chains, and marketing towards consumers.

Participants displayed a strong preference for the UNWW over the ICCO, suggesting that certifying agency plays a large role in the buyers' decision to pay a premium. Though we cannot know the reason the UNWW resulted in a higher probability of product consumption than the ICCO from our data, we offer several speculative explanations. Firstly, respondents may trust UNWW better than the ICCO for implementing a gender-equality programme. Moreover, by preferring a social justice NGO to a regulatory and quality-controlling entity, respondents may be indicating that they care less about the quality of their chocolate than about the social benefit their purchasing decision can make, when faced with two gender-equality labelled chocolate bars. Finally, the ICCO's purpose may not be as well known or as clearly understood from its name as the UNWW. We thus speculate that a gender-equality certification should be certified by a well known social justice and governance-related organization to ensure its success. Future studies would be of merit to ascertain the impact of candidates fitting that description, including the UNWW and Oxfam International (Oxfam, 2013), among other such institutions.

Finally, we observed a positive correlation between willingness to pay for our product and living in a community with high social capital, levels of activism and belief in ethical consumerism, as well as being female and a frequent buyer of dark chocolate. These results suggest that the bars may be most popular among individuals identifying as women in affluent, politically conscious and active areas. Areas like this might include cities like Washington DC and New York City. Both have high median incomes and are home to many NGOs and governmental organizations, suggesting a sizeable portion of the population comprises political and well-off individuals (Zou, 2016). Our findings suggest a healthy market is possible on US college campuses, where many students are from wealthier households than the national average and are often still supported by their parents, and may have a higher budget for casual spending (Golden, 2011; Chetty *et al.*, 2017). This information will be important to study for causality if the actual certification programme is employed, and gender-equality labelled chocolate bars enter the consumer market.

If marketed and promoted in such areas, with the aforementioned considerations about displayed attributes, such as location and structure of certification programme, the gender-equality labelled chocolate bar stands a chance of success as either a complement to an existing label, such as Fair Trade, or a standalone certification. Adeptly launched to meet market demand, a label denoting gender equality in cocoa production would be able to capitalize on the potential of women in agricultural development where existing labels have not. Interest in a hypothetical gender-equality label suggests promise in expanding the scope of humanitarian certification programmes to create a global market better suited to provide equal opportunity for all producers.

Notes

(1) Survey participants were presented with a hypothetical certification label and were told the labelling programme 'collects revenues from chocolate sales and corporate members' participation – when their product is granted the gender-equality label. The amounts generated are reallocated to agricultural and business training, microfinance and cooperative initiatives by and for women.'

(2) Intentional living communities are a housing solution offered to students by several universities as an alternative to dormitories. The otherwise named 'cooperative halls' in Cornell go above and beyond just sharing a house; they imply that the students who apply and are selected to live there will commit to contribute to community life through chores, cooking, house meetings and shared purchasing of food and supplies and services necessary to the communities. Although we cannot generalize to all intentional living communities, members of intentional living communities seem to have a higher propensity for social activism.

References and Further Reading

Andersen, J.G. and Tobiasen, M. (2004) Who are these political consumers anyway? Survey evidence for Denmark. In: Micheletti, M., Follesdal, A. and Stolle, D. (eds) *Politics, Products, and Markets: Exploring Political Consumerism Past and Present*. Transaction Publishers, New Brunswick, New Jersey, pp. 203–221.

Andorfer, V.A. and Liebe, U. (2012) Research on Fair Trade consumption – a review. *Journal of Business Ethics* 106(4), 415–435.

Andorfer, V.A. and Liebe, U. (2015) Do information, price, or morals influence ethical consumption? A natural field experiment and customer survey on the purchase of Fair Trade coffee. *Social Science Research* 52, 330–350.

Bacon, C.M. (2010) A spot of coffee in crisis: Nicaraguan smallholder cooperatives, Fair Trade networks, and gendered empowerment. *Latin American Perspectives* 37(2), 50–71.

Basu, A.K. and Hicks, R.L. (2008) Label performance and the willingness to pay for fair-trade coffee: a cross-national perspective. *International Journal of Consumer Studies* 32(5), 470–479.

Boersma, F.V. (2009) The urgency and necessity of a different type of market: the perspective of producers organized within the Fair Trade market. *Journal of Business Ethics* 86(1), 51–61.

Boivin, C., Parissier, C., Lapointe, S., Emond, S., Sicotte, A. and Theoret Legault, H. (2012) What are consumers looking for in dark chocolate? *Journal of Business & Economics Research* 10(3).

Buvinic, M., Das Gupta, M. and Casabone, U. (2009) Gender, poverty and demography: an overview. *The World Bank Economic Review* 23(3), 347–369.

Chan, M. and Barrientos, S. (2010) *Improving Opportunities for Women in Smallholder-Based Supply Chains*. Bill & Melinda Gates Foundation, Seattle, Washington.

Chetty, R., Friedman, J.N., Saez, E., Turner, N. and Yagan, D. (2017) *Mobility Report Cards: the Role of Colleges in Intergenerational Mobility*. Equality of Opportunity Project. Available at: http://www.equality-of-opportunity.org (accessed 5 July 2017).

Cook, J. (2008) *Commodity Prices Update: Coffee, Cocoa, and Cotton*. University of Washington, Seattle, Washington.

Cornell University (2016) *University fact: Cornell by the numbers*. Available at: https://www.cornell.edu/about/facts/cfm (accessed 10 February 2016).

Cunningham, C.E., Deal, K. and Chen, Y. (2010) Adaptive choice-based conjoint analysis: a new patient-centred approach to the assessment of health service preferences. *The Patient* 3(4), 257–273.

Doran, C.J. (2010) Fair trade consumption: in support of the out-group. *Journal of Business Ethics* 95(4), 527–541.

Fairtrade International (2015) *Equal Harvest: Removing Barriers to Women's Participation in Smallholder Agriculture*. Fairtrade International, Bonn, Germany.

FAO (2011) *The State of Food and Agriculture: Women in Agriculture Closing the Gender Gap for Development*. Food and Agriculture Organization of the United Nations, Rome.

Forno, F. and Ceccarini, L. (2006) From the street to the shops: the rise of new forms of political actions in Italy. *South European Society & Politics* 11, 197–222.

Golden, P. (2011) *Nearly 60 Percent of Parents Financially Supporting Adult Children*. National Endowment for Financial Education, Denver, Colorado.

Gonzalez, B.M. and Macleod, M. (2010) Challenging gender inequality in farmers' organizations in Nicaragua. *Gender and Development* 18(3), 373–386.

Green, P.E. and Srinivasanm, V. (1990) Conjoint analysis in marketing: new development with implications for research and practice. *Journal of Marketing* 54(4), 3–19.

Greig, D. (2006) Shea butter: connecting rural Burkinabé women to international markets through Fair Trade. *Development in Practice* 16(5), 465–475.

Haddon, M. and Rai, N. (2012) Chocolate's recession-proof reputation is tested. *The Wall Street Journal*. Available at: http://www.wsj.com/articles/SB10000872396390443675404578058103684551128 (accessed 5 July 2017).

Hainmueller, J., Hiscox, M.J. and Sequeira, S. (2014) Consumer demand for the Fair Trade label: evidence from a multi-store field experiment. *Review of Economics and Statistics* 97(2), 242–256.

Hickey, W. (2014) The Dollar-and-Cents case against Hollywood's exclusion of women. *FiveThirtyEight*. Available at: http://fivethirtyeight.com/features/the-dollar-and-cents-case-against-hollywoods-exclusion-of-women (accessed 5 February 2016).

Hutchens, A. (2010) Empowering women through fair trade? Lessons from Asia. *Third World Quarterly* 31(3), 449–467.

IIPS (2007) *National Family Health Survey (NFHS-3): 2005–06: India*. International Institute for Population Sciences, Mumbai.

Laven, A. (2012) *Women & Cocoa Farming: 'bridging the gap'*. World Cocoa Foundation Partnership, Zurich.

Lyon, S. (2008) We want to be equal to them: fair trade coffee certification and gender equity within organizations. *Human Organization* 67(3), 258–268.

Malik, K. (2013a) *Human Development Report 2013: Honduras*. United Nations Development Programme, New York.

Malik, K. (2013b) *Human Development Report 2013: Cote d'Ivoire*. United Nations Development Programme, New York.

Malik, K. (2013c) *Human Development Report 2013: Indonesia*. New York: United Nations Development Programme, New York.

Malik, K. (2013d) *Human Development Report 2013: The Rise of the South*. United Nations Development Programme, New York.

Menon, N., Van Der Meulen Rogers, Y. and Kennedy, A.R. (2016) Land reform and welfare in Vietnam: why gender of the land-rights holder matters. *Journal of International Development* 29(4), 454–472.

Neilson, L.A. and Paxton, P. (2010) Social capital and political consumerism: a multilevel analysis. *Social Problems* 57(1), 5–24.

Newman, B.J. and Bartels, B.L. (2011) Politics at the checkout line: explaining political consumerism in the United States. *Political Research Quarterly* 64(4), 803–817.

Noland, M., Moran, T. and Kotschwar, B. (2016) *Is Gender Diversity Profitable? Evidence from a Global Survey*. Working Paper 16-3. Peterson Institute for International Economics, Washington, DC.

Oxfam (2013) *Equality for Women Starts with Chocolate*. GROW Campaign, Oxfam International, Oxford. Available at: http://www.oxfam.org/sites/www.oxfam.org/files/gender-inequality-cocoa-nigeria.pdf (accessed 5 July 2017).

Pew Research Center (2015) *Global Attitudes Survey* A56c. [Dataset and analysis] Available at: http://www.pewglobal.org/2015/11/18/1-support-for-democratic-principles/ (accessed 5 February 2016).

Quisumbing, A.R., Mainzen-Dick, R., Raney, T.L., Croppenstedt, A., Behrman, J.A. and Peterman, A. (eds) (2014) *Gender in Agriculture: Closing the Knowledge Gap*. International Food Policy Research Institute (IFPRI), Washington, DC.

Shah, D.V., McLeod, D.M., Kim, E., Lee, S.Y., Gotlieb, M.R., Ho, S.S. and Breivik, H. (2007) Political consumerism: how communication and consumption orientations drive 'Lifestyle politics'. *Annals of the American Academy of Political and Social Science* 611, 217–234.

Stolle, D., Hooghe, M. and Micheletti, M. (2005) Politics in the supermarket: political consumerism as a form of political participation. *International Political Science Review* 26(3), 245–269.

Tallontire, A., Dolan, C., Smith, S. and Barrientos, S. (2005) Reaching the marginalised? Gender value chains and ethical trade in African horticulture. *Development in Practice* 15(3/4), 559–571.

UN Women (2015) *Gender equality is a 'Silver Bullet' for development: Helen Clark*. UN Women Regional Office for Asia and the Pacific, Bangkok. Available at: http://asiapacific.unwomen.org/en/news-and-events/stories/2015/12/gender-equality-is-a-silver-bullet-for-development (accessed 5 February 2016).

UTZ (2009) *The Role of Certification and Producer Support in Promoting Gender Equality in Cocoa Production*. UTZ Certified, Solidaridad-Certification Support Network (CSN), Amsterdam, in cooperation with Oxfam Novib.

Valkila, J., Haaparanta, P. and Niemi, N. (2010) Empowering coffee traders? The coffee value chain from Nicaraguan fair trade farmers to Finnish consumers. *Journal of Business Ethics* 97(2), 257–270.

Zou, M. (2016) The DC area has the highest median income in the US again. *The Washingtonian*. Available at: https://www.washingtonian.com/2016/09/21/the-dc-area-has-the-highest-median-income-in-the-us-again/ (accessed 10 October 2016).

10 Fair Trade Certification on Plantations: Household Wealth and Welfare Implications for Hired Labour

Katharina Krumbiegel* and Meike Wollni
[1]*University of Goettingen, Department of Agricultural Economics and Rural Development, Germany*

10.1 Introduction

10.1.1 Fair Trade certified plantation agriculture

About 1.3 billion workers are employed in the agricultural sector worldwide, of which about 500 million work as casual, temporary or permanent workers on plantations. Hired labour on plantations or in factories are considered one of the most vulnerable groups in the global trade system. They are often exposed to discrimination, difficult working conditions, low wages and lack of bargaining opportunities. In recent years, however, consumers have become increasingly aware of unfavourable employment conditions in the food producing industry. This awareness has been mirrored by the rise of private food and sustainability standards, such as Fair Trade. While the Fair Trade movement originally aimed to empower small-scale farmers to overcome global trade barriers of limited price information, organization and production in a free market, its support was further extended to plantation workers in the 1990s. Fair Trade's aim is 'to empower workers and the sustainable social and economic development of workers and their communities' (Fairtrade International, 2014). In this context, the Fair Trade certification of large-scale agricultural companies aims to improve not only working conditions but also socioeconomic well-being of workers and their communities. While some studies have focused on the evaluation of general certification effects (particularly of GlobalGAP) in the employment context, very little is still known about the specific role of Fair Trade for workers' household wealth accumulation and welfare benefits.

Numerous studies and evaluations assess Fair Trade's impact on small-scale farmers regarding income, farm productivity and poverty reduction, but only a few studies look at the implications of Fair Trade certification on plantations for workers and their households (Lyall, 2014). Granville and Telford (2013) found via descriptive comparison that Fair Trade workers earned salaries above the minimum wage in the wine industry in South Africa. Ruben and van Schendel (2009) compared 50 workers on one Fair Trade certified banana plantation to a non-certified one. They found that workers on the non-certified plantation received a higher salary, but also worked longer hours and received smaller benefits (Dragusanu *et al.*, 2014). A study by the 'Fair Trade,

*Corresponding author; e-mail: katharina.krumbiegel@agr.uni-goettingen.de

Employment and Poverty Reduction' project from the University of London did not find evidence for higher wages through Fair Trade certification on small farms and large estate units in the tea, coffee and flower sectors in Uganda and Ethiopia (Cramer *et al.*, 2014). Raynolds (2012) found that Fair Trade's benefits for workers particularly lay in its ability to empower them and secure their well-being at work. The empowerment pathway was mainly driven by Fair Trade-mandated worker committees in the Ecuadorian flower sector, where labour unions were largely absent.

While Fair Trade certification of large-scale plantations is predominantly concerned with working conditions, wages and worker empowerment, ultimately Fair Trade aims to create a framework that allows workers to provide education for their children, ensure household food security and reduce poverty. There are potential channels through which wealth and living standards may be enhanced for workers on Fair Trade certified plantations. Firstly, Fair Trade products receive a guaranteed minimum price in the Fair Trade market, which is not only to cover the costs of sustainable production but also to meet a so-called living wage within the particular sector. This guaranteed price could contribute to potentially higher income levels going beyond the minimum wage, and thus may provide consistent income for the worker's household. Secondly, consistency could further be supported by stable work arrangements. Work contracts provided on Fair Trade plantations may be more permanent, contributing to a household's ability to buffer shocks and increase its resilience. Thirdly, certain working conditions such as payment of social security, loan availability and health care provisions may ease household costs and provide extra funds for investments in education, asset accumulation or agricultural inputs. Fourthly, a Fair Trade premium that is paid on top of the guaranteed market floor price is invested in community infrastructure, vocational trainings or educational projects. This may raise overall community welfare and therefore also indirectly affect workers' household welfare levels. Against this background, this study aims to address the following research questions:

- Does Fair Trade positively affect the income level of workers' households?
- Does Fair Trade contribute to wealth accumulation and higher living standards?

10.1.2 The export pineapple sector in Ghana

The case study presented in this chapter is based on cross-sectional data from hired labour in the Ghanaian pineapple sector. Pineapple is one of Ghana's most important horticultural export crops. Its production was introduced to Ghana by smallholder farmers in the 1980s. With a rising demand, particularly from Europe, large-scale farmers began acquiring land for export production close to shipping ports and airports (Fold and Gough, 2008). The dominant pineapple type exported was 'Smooth Cayenne', a variety that is adapted to local growing conditions. In the 1990s, Ghana was the third most important pineapple supplier to the European Union after Côte d'Ivoire and Costa Rica. In the late 1990s, the company Fresh Del Monte developed a new pineapple variety called 'MD2', the so-called 'shipping pineapple' with a high level of sweetness, low acidity and longer shelf-life than previous varieties. 'MD2' is considered an industrial crop that has been developed for large-scale and more mechanized production. Its expansion in Costa Rica coupled with vast marketing campaigns in the USA and Europe ultimately changed consumer taste in favour of the new variety. At the same time, production levels increased in multiple countries and eventually led to falling export prices. Because 'MD2' requires fertilizer, pesticides, plastic mulching and cooling facilities, it demands high initial and continuous capital investments. Ghanaian producers were unable to adapt to the quick change, due to information and capital constraints, leading to a decline in the EU market share from 10.5% in 2003 to 4.3% in 2007 (Fold and Gough, 2008; Kleemann *et al.*, 2014; Harou *et al.*, 2017). This has forced the majority of small-scale farmers to drop out of export-oriented production and today they predominantly sell to the local market or to processors. Large-scale farms were better able to adjust to the changes taking place in the industry. Today, about 15 large-scale plantations produce pineapples for the export market, eight of which make up 93% of Ghana's pineapple export volume (Gatune *et al.*, 2013). In 2011, Ghana's export value of fresh and processed pineapple was worth $51 million, representing the country's sixth most important export crop in terms of value (Gatune *et al.*, 2013). All of the exporting plantations are GlobalGAP-certified, which constitutes a minimum requirement to export to the European Union, the main market for Ghanaian pineapple. Additionally, about half of the plantations are Fair Trade certified. Fierce competition in international pineapple markets has motivated many Ghanaian plantations to target this particular niche market. Overall, increasing demand for quality assurance, consistent supply and certification remain challenging for Ghanaian producers.

10.2 Data and Variables

10.2.1 Survey and sample

Our dataset consists of original survey data from 361 hired plantation workers and their households living in rural areas in the Ghanaian pineapple belt. This central area for pineapple production stretches across four different regions: Central Region, Eastern Region, Greater Accra Region and Volta. For the purpose of our study, we restricted our sample to manual or low-skilled labourers and therefore excluded the management, administrative and technical levels, resulting in a sample of 325 observations for the analysis. The data collection process incorporated two stages. In November 2014 we held semi-structured interviews with main stakeholders in the pineapple export sector, including representatives from agricultural ministerial divisions at the central and district level, the association of sea-freight pineapple exporters of Ghana, foreign aid agencies, and management boards from pineapple producing and processing companies. In the second stage, we collected original survey data from workers between April and July 2015. We purposefully selected four Fair Trade and four non-Fair Trade certified companies that were comparable to each other. We compared our companies based on information regarding: (i) size in area; (ii) size in workers; (iii) production capacity levels in metric tons (t) per week; and (iv) involvement of foreign management. Table 10.1 shows that Fair Trade companies work on a larger scale than non-Fair Trade companies, particularly in terms of the area of production and the workers employed. On average, the Fair Trade companies had been Fair Trade certified for 3.63 years; individually, however, this ranged from 2 to 14 years. On average, Fair Trade companies sold approximately 38% of their production into the Fair Trade market.

From each company, we obtained lists of workers in the villages from which the companies source their labourers. Based on these lists, we drew a stratified random sample of workers by company for the interviews. The structured questionnaire incorporated questions on household characteristics, family health and dietary diversity, land ownership and agricultural production as well as employment conditions, provision of services, labour union involvement and social projects implemented by the companies. Interviews were conducted face-to-face by local field assistants with an agricultural economics background and who participated in in-depth questionnaire training and pre-testing. Most of the households in our sample comprised a household head (usually the husband) and a

Table 10.1. Overview of the companies.

Variable *N* = 8	Fair Trade certified company: Mean value	Fair Trade certified company: Standard deviation	Non-Fair Trade certified company: Mean value	Non-Fair Trade certified company: Standard deviation	Difference and test statistics
Size of the company (ha)	338	122.32	190	58.31	148*
Size of the company (worker numbers)	347.50	112.66	148.50	48.12	199**
Productivity level (t/week)	165	107.55	79	55.53	86
Foreign involvement in company management[a]	0.75	0.5	0.75	0.5	0

[a]1 = at least one person on company's management board has foreign nationality; 0 = all persons on management board have Ghanaian nationality. *Significant at 10%; **significant at 5%; ***significant at 1%.

spouse. When there was no husband, we considered the female to be the household head. In our scenario of plantation employment, either the household head or the spouse was interviewed as a worker of a pineapple plantation. In some cases the husband/wife also worked for the same company. In our analysis, we differentiated between Fair Trade workers on Fair Trade certified plantations (and their households) and workers on non-Fair Trade certified plantations (and their households). For simplicity, we refer to them as Fair Trade workers (FT workers) and non-Fair Trade workers (non-FT workers).

10.2.2 Measuring welfare indicators

In the context of this study, we are interested in evaluating the impact of wage labour on Fair Trade certified plantations based on several welfare indicators. Firstly, we consider the contribution of wage labour income to overall household income. Household income is aggregated for the past 12 months and includes income from selling agricultural produce, income from own business, income from off-farm employment as well as income from other sources, such as pensions, gifts and remittances. Secondly, we are interested in asset accumulation as an indicator for more long-term wealth accumulation. In this assessment we closely follow Filmer and Pritchett (2001) and construct a standardized asset index via principal component analysis. Principal components enable us to capture the most common information from linear combinations of the asset variables (Filmer and Pritchett, 2001). In the composition of the index we include 13 variables[1] that reflect the possession of the following assets: motor vehicle, motorbike, bicycle, fan, freezer, sewing machine, water tank, gas stove, jewellery, kente cloth, TV set, radio and bank account. For easier interpretation we transform the values into a standardized asset index that is calibrated on a 0–100 scale by the formula used in Sekhar *et al.* (1991).

Thirdly, we are interested in looking into different proxy indicators for standard of living. In particular, we consider access to clean drinking water and electricity. By including provisions predominantly made at the village level into our analysis, we address more overarching development indicators. Apart from Fair Trade's potential impact on household income, Fair Trade certification of companies may positively affect infrastructure provision at the village level. The Fair Trade premium is spent on social and infrastructural projects with both workers and villages as beneficiaries, such as building groundwater pumps or toilet facilities. Furthermore, the presence of export-oriented companies overall may improve village settings due to their demand for water, electricity, a sewage system, etc. In particular, Fair Trade companies may also use their bargaining power with local governments to have these services provided, as they might be more interested in ensuring village services surrounding their company grounds. In the analysis we also control for village characteristics and infrastructure in addition to household level characteristics (income, education, etc.). Drinking water is defined as clean when derived from improved sources: household connection, public standpipe, borehole or pump, protected dug well, protected spring, rainwater collection (WHO and UNICEF, 2015). Rivers, lakes and streams as well as unprotected wells and springs are considered unimproved sources of drinking water (WHO and UNICEF, 2015). Access to electricity is measured as the main source of lighting.

10.3 Descriptive Statistics

Household demographics are presented in Table 10.2. Fair Trade workers are on average 2.4 years older and have more dependents (children below the age of 18 and/or adults above the age of 65) living in the household than non-Fair Trade workers. The distribution of female-headed households and religious affiliation are similar between the two comparison groups. Non-Fair Trade workers seem to have slightly better education levels with a higher number of workers being at least secondary school graduates and a lower share with no formal education at all. Literacy rates are nonetheless comparable across all workers. When looking at the living conditions indicators, we see that households of FT workers are more likely to have electricity as well as access to clean drinking water. In addition, the composite asset index shows that FT workers have a higher number of assets in comparison with non-FT workers.

Table 10.2. Summary statistics of worker and household demographics including indicators of wealth and standard of living.

Variable	Household with at least one Fair Trade worker		Household with at least one non-Fair Trade worker		Difference and test statistics
	Mean value	Standard deviation	Mean value	Standard deviation	
Household Demographics					
Number of workers in household	1.49	0.61	1.36	0.49	0.14**
Number of workers on pineapple plantations in household	1.23	0.47	1.10	0.30	0.13***
Number of dependents	2.52	1.54	1.97	1.43	0.55***
Female household head	0.23		0.28		0.04
Protestant	0.85		0.86		0.01
Catholic	0.03		0.06		0.03
Muslim	0.04		0.02		0.02
Worker Demographics					
Female worker	0.62		0.61		0.01
Worker is married	0.81		0.67		0.13***
Worker is literate	0.46		0.50		0.04
Worker did not go to school	0.33		0.15		0.17***
Worker finished primary school only	0.23		0.23		0.01
Worker finished secondary school or higher	0.44		0.62		0.18***
Age of worker (years)	38.51	9.10	36.07	10.46	2.44**
Living conditions					
Electricity	0.86		0.78		0.08*
Earthen floor	0.17		0.11		0.06
Access to improved sanitation	0.24		0.19		0.06
Clean drinking water	0.88		0.69		0.19***
Total agricultural land	1.02	1.35	0.98	1.22	0. 04
Standardized Asset Index	22.47	17.45	15.77	16.95	6.70***

*Significant at 10%; **significant at 5%; ***significant at 1%.

In Table 10.3 we take a closer look at income-generating activities and company provisions contributing to total household income and welfare. Concerning households' income levels, we differentiate between: (i) income from horticultural wage employment; (ii) income generated on own agricultural land; (iii) income from self-employment (such as tailoring, shop keeping, hair dressing); (iv) income from off-farm wage employment; and (v) additional income from pensions, gifts and others. While most income sources are equally distributed across the two different groups, the income from horticultural wage labour for Fair Trade worker households is significantly higher than for non-Fair Trade worker households. This may in part be explained by higher hourly wages paid to FT workers. The vast majority of FT workers are also hired under a permanent work contract. A permanent employment status provides worker households with more security for long-term planning and, combined with access to credit, potentially enables them to make long-term investments. FT workers also participate in more training and are more likely to be engaged in a labour union.

10.4 Methodology

10.4.1 Regression analysis

To analyse the impact of wage labour on Fair Trade certified plantations on household wage labour income, we estimate the following linear regression model:

Table 10.3. Summary statistics of worker incomes and company provisions.

Variable	Household with at least one Fair Trade worker		Household with at least one non-Fair Trade worker		Difference and test statistics
	Mean	Standard deviation	Mean	Standard deviation	
Household incomes (in GHS)[a]					
Total income	5720.05	5951.88	4068.02	3272.86	1652.03***
(1) Horticultural wage labour income	3116.36	1339.32	2602.60	1101.19	513.76***
(2) Agricultural income	960.99	3734.57	559.20	2808.91	401.79
(3) Self-employment income	720.98	1580.82	584.78	1449.81	136.20
(4) Other wage labour income	573.25	2441.30	264.60	1306.67	308.65
(5) Other income	75.12	287.77	56.82	155.66	18.30
Workplace conditions					
Hourly salary (GHS)[a]	1.54	1.39	1.17	0.61	0.38***
Daily salary (GHS)[a]	10.10	6.02	9.22	4.18	1.73***
Permanent employment	0.87		0.53		0.34***
Labour union membership (if a labour union is present at the company)	0.73		0.45		0.27***
Number of trainings received in past 12 months	1.70	2.86	0.40	1.44	1.30***

[a]Ghana cedi (¢); *significant at 10%; **significant at 5%; ***significant at 1%.

$$\textit{wage labour income}_i = \alpha_0 + \alpha_1 FT_i + \alpha_2 x_i + \varepsilon_i \quad (10.1)$$

where wage labour income is measured as the total income earned from horticultural wage labour during the past 12 months by household i; FT indicates whether the workers are employed on a Fair Trade certified plantation; x is a vector of household, worker and company specific characteristics; and ε is a random error term. The treatment dummy FT takes a value of one if the pineapple plantation is Fair Trade certified and zero otherwise. Regarding company-specific variables, we take productivity levels, company size (in hectares) and employment conditions into account. Furthermore, we include worker demographics and worker ability (whether the worker grows pineapple on their own farm and whether they have worked for a fruit company before taking up the current job). Additionally, we control for the different types of manual labour jobs. Finally, given that there might be more than one worker working on a pineapple plantation in the household, we control for the number of workers in the household.

Secondly, we estimate a linear regression model to investigate the effect of Fair Trade certification on asset accumulation:

$$\textit{standardized asset index}_i = \beta_0 + \beta_1 FT_i + \beta_2 x_i + \varepsilon_i \quad (10.2)$$

where the standardized asset index for household i is derived from principal component analysis and ranges between zero (low asset accumulation) and 100 (high asset accumulation). Vector x includes household demographics expected to influence the purchase and accumulation of assets at the household level.

Thirdly, we estimate a probit model to evaluate the impact of Fair Trade certification on two selected standard-of-living indicators (access to clean drinking water and electricity):

$$\textit{standard of living}_i = \gamma_0 + \gamma_1 FT_i + \gamma_2 x_i + \varepsilon_i \quad (10.3)$$

where the standard of living for household i is specified as one of two dummy variables taking the value one if the household has access to clean drinking water or to electricity, respectively. The vector x incorporates household- as well as village-specific characteristics, such as infrastructure and service availability. To account for the fact that households are more comparable on the village level, we cluster the standard errors at village level for all regressions.

10.4.2 Sample restriction and propensity score matching

As individuals can choose to some extent whether they work for a Fair Trade certified or a non-Fair Trade certified company, there may be unobserved characteristics, such as motivation or dedication, that influence the choice of work place as well as the outcome variables. This could potentially lead to selection bias in the estimation of the linear regression and probit models introduced in the previous section. Effectively, the freedom to choose a work place in rural Ghana is often restricted due to long distances and limited access to transport. Therefore, in practice, workers select their work place mainly based on proximity to their village to reduce daily commute and, vice versa, companies source the majority of their workers from villages surrounding their estate units. This is also reflected in our data. Our workers come from 56 villages. In 29 villages there are only FT workers and in 25 villages there are only non-FT workers. In only two of these villages are both FT and non-FT workers located. This shows that indeed workers do not necessarily work for the company they consider 'best' (which may be correlated with Fair Trade) but base their decision to take up work on the proximity and access to the closest company. Further, medium-term migration for paid labour also seems to be relatively limited in comparison with other export-oriented production sites. From our sample, only 16% of all workers state that they have resettled to the pineapple growing area for work reasons. Still, to reduce potential selection bias, we apply a sample restriction based on the propensity score as well as a matching approach (Rosenbaum and Rubin, 1983). This enables us to compare and match FT and non-FT workers based on certain observable characteristics, which are assumed to be correlated with the unobservable characteristics.(2)

We define treatment FT as a binary variable that equals one if the worker in the household is employed by a Fair Trade company. Using a vector of observed variables (x), we then predict the probability of working on a Fair Trade plantation (the propensity score) to create a comparable counterfactual group:

$$p(x) = \Pr\{T = 1|x\} = E\{T|x\} \quad (10.4)$$

We select covariates that satisfy the Conditional Independence Assumption (CIA) by considering those that affect participation. A probit model is used to regress the binary treatment variable on worker characteristics and derive the propensity score. Table 10.4 shows the

Table 10.4. Determining propensity scores. Probit model of Fair Trade workers.

Variable	Marginal effects	Standard error
Age of worker	0.136**	0.06
Age of worker^2	–0.002**	0.001
Female household head	0.525*	0.298
Worker is married	0.892***	0.320
Worker has no school education	0.754***	0.197
Worker has primary school education	0.210	0.186
Household size	0.178**	0.91
Number of dependents	–0.064	0.107
Household head was born in village where lives today	–0.264*	0.155
Protestant	–0.149	0.263
Catholic	–0.409	0.456
Agricultural land size under production	–0.353**	0.156
Agricultural land size under production^2	0.063*	0.035
Observations	322	
Pseudo R^2	0.13	

*Significant at 10%; **significant at 5%; ***significant at 1%.

overview of covariates included in the probit model. The region of common support is between 0.10 and 0.98 and the balancing property is satisfied. Based on the estimated propensity scores, we trim the observations outside of the area of common support and estimate the regression specifications introduced above using only those observations that fall within the area of common support.

As robustness check to our regression analyses, we apply propensity score matching techniques. Based on the matched households, we calculate the average treatment effect on the treated (ATT) by comparing the outcomes between the treated and the control group:

$$ATT = E(\Delta | p(x), T = 1) = E(y_1 | p(x), T = 1) - E(y_0 | p(x), T = 1) \quad (10.5)$$

We apply three different matching techniques with bootstrapped standard errors (Caliendo and Kopeinig, 2005): (i) kernel matching (using weighted averages of all individuals in the control group to construct a counterfactual outcome); (ii) radius matching (imposing a tolerance level on the maximum propensity score distance (caliper, here = 0.1)); and (iii) stratification matching (partitioning the common support of the propensity score into a set of intervals and calculating the impact within each interval by taking the mean difference in outcomes between treated and control observations). Results are presented in Table 10.5. The propensity score matching results confirm our regression findings.

10.5 Findings

10.5.1 Fair Trade certification and horticultural wage labour income

Results on the determinants of horticultural wage labour income are presented in Table 10.6. We find that Fair Trade certification contributes significantly to higher horticultural wage labour income, increasing average yearly incomes by approximately 560–580 Ghana cedi (¢).[3] This can be attributed to higher wages on Fair Trade certified pineapple plantations (see descriptive statistics). With respect to the company-specific variables, we find that company size has a significantly negative effect on wage labour income, indicating that smaller plantations may provide better wage conditions. The size of the effect, however, is very small. Furthermore, company productivity also matters. Higher output positively influences horticultural wage labour income, albeit the magnitude of the effect is very small as well. Regarding household-specific characteristics, as expected the number of persons

Table 10.5. Matching results for Fair Trade workers for horticultural wage labour income, asset index, access to clean drinking water and electricity.

	Average treatment effect on the treated	Standard error	*t*-value	Treated (no.)	Controlled (no.)
Horticultural wage labour income					
Kernel matching	670.67	144.45	4.64	164	152
Radius matching (with 0.1 caliper)	542.04	149.03	3.64	164	152
Stratification matching	680.62	159.78	4.26	164	152
Asset Index					
Kernel matching	6.46	1.68	3.86	164	152
Radius matching (with 0.1 caliper)	5.29	2.21	2.40	164	152
Stratification matching	6.93	2.10	3.31	164	152
Access to clean drinking water					
Kernel matching	0.16	0.04	3.70	164	152
Radius matching (with 0.1 caliper)	0.19	0.05	3.82	164	152
Stratification matching	0.18	0.05	3.34	164	152
Access to electricity					
Kernel matching	0.10	0.05	1.81	164	152
Radius matching (with 0.1 caliper)	0.11	0.05	2.37	164	152
Stratification matching	0.10	0.05	1.90	164	152

employed on horticultural plantations increases horticultural wage income at the household level. Female workers earn almost ₵200 less than their male co-workers per year. This may be due to the different kind of jobs men and women do, the former potentially engaged in more qualified tasks or in more supervisory roles within the sectors we control for. Indicators of experience in pineapple production do not determine wage labour income: neither having previously worked for a fruit company nor growing pineapple on their own farm influences horticultural wage income.

Table 10.6. Regression results for horticultural wage labour income of hired labour.

Variable	OLS[a]	OLS with restricted sample
Fair Trade certification	589.2**	567.5**
	(256.5)	(261.4)
Size of company (in used ha)	−3.167***	−3.188***
	(1.097)	(1.104)
Productivity of company (t/week)	1.207	1.397
	(1.226)	(1.221)
Worker is female	−197.6*	−193.9*
	(103.7)	(105.5)
Worker finished secondary school	173.7	161.8
	(138.9)	(138.0)
Age of worker	−9.160	−9.492
	(6.470)	(6.862)
Number of workers in household	2,055***	2,057***
	(156.1)	(154.3)
Previously worked for a fruit company	174.1	182.7
	(178.9)	(190.3)
Grows pineapple on own farm	133.5	160.1
	(176.4)	(171.5)
Works in planting-related activities	−35.98	−31.60
	(140.3)	(147.2)
Works in export-related activities	131.9	104.3
	(148.3)	(153.1)
Works with chemical application	83.17	85.26
	(163.0)	(168.5)
Works in sucker management	92.94	87.05
	(181.7)	(183.9)
Works in other menial tasks	425.1**	489.0**
	(210.8)	(211.3)
Constant	1,122**	1,128**
	(488.0)	(509.0)
N	325	315
R^2	0.551	0.562

[a]Ordinary least square; *significant at 10%; **significant at 5%; ***significant at 1%. Standard errors in parentheses.

10.5.2 Fair Trade certification and asset accumulation

We are now interested in the role of Fair Trade certification on asset accumulation, which is commonly regarded as a more durable indicator than income. In Table 10.7 we see that households with FT worker(s) have a significantly higher asset index than households with non-FT workers. Interestingly, we further see that particularly female-headed households have much higher asset accumulation ratios than male-headed households. A reason for this could potentially be that women are restricted in other wealth accumulation strategies, such as buying land, and therefore have to resort to asset accumulation in particular. Marriage, higher education levels and more agricultural land all highly correlate with increased accumulation of assets, as is expected.

Table 10.7. Regression results for asset index of workers' households.

Variable	OLS[a]	OLS with restricted sample
Worker(s) in household works for a Fair Trade certified pineapple company	7.793***	7.263***
	(2.750)	(2.016)
Household head is female	9.207*	9.898***
	(4.634)	(3.700)
Age of household head	0.041	0.002
	(0.069)	(0.10)
Household head is married	10.58**	10.79***
	(4.466)	(3.838)
Household head finished primary school only	5.198**	4.950*
	(2.053)	(2.995)
Household head finished secondary school or higher	12.79***	12.59***
	(2.181)	(2.566)
Number of dependents	−0.191	−0.209
	(0.602)	(0.641)
Agricultural land used for production	1.870*	1.972**
	(1.090)	(0.763)
Distance to market	−0.301**	−0.311**
	(0.126)	(0.158)
Constant	−3.255	−1.552
	(5.741)	(5.929)
N	325	315
R^2	0.192	0.187

[a]Ordinary least square; *significant at 10%; **significant at 5%; ***significant at 1%. Standard errors in parentheses.

One potential impact channel through which Fair Trade leads to higher asset accumulation may be through the income pathway. Increases in horticultural wage labour income may be of particular importance, given that it represents the major source of income for the households in our sample, contributing approximately 60% to total household income. However, Fair Trade certification may affect asset accumulation not only through the income pathway, but through other channels as well. Work arrangements are more stable and permanent on Fair Trade certified plantations and can therefore contribute to more long-term decision making when it comes to investments. Fair Trade may also ease expenditure constraints through payments of health care needs and social security contributions (Table 10.8) as well as social and economic projects that are funded by the Fair Trade premium. Fair Trade premium-funded social and economic projects targeting the individual level can range from payment of school fees for secondary school children of workers, provision of uniforms and/or books to school-aged children of workers to micro-finance loans with lower interest rates than those of official banks and lenders. We can see in Table 10.8 that 24% of the Fair Trade workers have used a loan that was facilitated by their company. Access to loans enables investments in assets. In contrast, only 5% of non-Fair Trade workers were able to do the same. These may be additional pathways of being able to accumulate assets for workers' households.

10.5.3 Fair Trade certification and standard of living indicators

Finally, Tables 10.9 and 10.10 present the results on two proxy indicators for standard of living: access to clean drinking water and access to electricity. We find that Fair Trade certification positively correlates with both indicators. This could be due to direct effects from higher horticultural wage labour income at the household level but also to more indirect effects, such as company presence or their social projects at the village level. Given that both drinking water and electricity are provided on the village level, village-specific indicators play an important explanatory role in comparison with household characteristics. The level of development, location and access to community services all correlate strongly with the two standards of living indicators. The more development projects[(4)] there are in a village, the more likely it is that villagers have access to clean drinking water and electricity. This access becomes more difficult the farther the village is situated from a tarred road, particularly for electricity provision. Interestingly the distance from a horticultural company is positively correlated with access to clean drinking water. The reason for households in proximity to companies to have reduced access to clean drinking water may derive from the definition of clean drinking water. Pineapple companies usually set up production facilities close to surface water bodies such as rivers and lakes for the irrigation of their crops. Sometimes they also build water reservoirs to cater to their needs. Villages located close to the companies may indeed use the same water sources. By definition, these surface water sources are unimproved.

As mentioned before, standard of living indicators, such as access to clean water and electricity, are predominantly provided on the village level and can be a direct result of projects financed by the Fair Trade premium. Table 10.11 shows exemplary village-based projects and beneficiaries. While this list is by no means complete, it gives an indication of potential channels of Fair Trade contributing to village-level welfare outcomes.

In general, such projects may certainly also be implemented by non-Fair Trade certified companies, but this is not the case in our study area. In general these projects are appreciated by worker's households. More than half of the projects mentioned are stated to benefit the households of the workers interviewed. Even though some projects cater to fewer beneficiaries (e.g. families with children or young mothers), the vast majority of workers (almost 70%) prefer the implementation of the projects instead of just an equal distribution of money. This is contrary to what is suggested during interviews with company managements, who feel that projects are very often not appreciated by the workers if they do not directly benefit from them. Combining the more quantitative findings of Tables 10.9 and 10.10, which show a positive correlation between Fair Trade certification and higher living standards, with the subjective statements of the workers'

Table 10.8. Provision of services within companies.

Variable	Household with at least one Fair Trade worker: Mean value	Household with at least one Fair Trade worker: Standard deviation	Household with at least one non-Fair Trade worker: Mean value	Household with at least one non-Fair Trade worker: Standard deviation	Difference and test statistics
Company services used					
Lunch	0.21	0.41	0.28	0.45	0.07
Transport	0.49	0.50	0.70	0.46	0.21***
Medical care for worker on site	0.64	0.48	0.35	0.48	0.29***
Medical care for worker off site	0.59	0.49	0.30	0.49	0.19***
Medical care for family off site	0.06	0.24	0.01	0.08	0.05***
Social allowances (for funerals etc.)	0.07	0.25	0.01	0.11	0.05*
Loan	0.24	0.43	0.05	0.22	0.19***

*Significant at 10%; **significant at 5%; ***significant at 1%.

Table 10.9. Probit results for standard of living indicator 1: access to clean drinking water.

Variable	Probit	Probit with restricted sample
Worker(s) in household works for a Fair Trade certified pineapple company	0.788***	0.885***
	(0.244)	(0.280)
Income (in 1000 GHS)[a]	0.009	0.008
	(0.020)	(0.020)
Household head is female	−0.012	−0.166
	(0.282)	(0.301)
Age of household head	0.027***	0.032***
	(0.010)	(0.009)
Household head is married	−0.324	−0.619*
	(0.295)	(0.321)
Household head finished primary school only	−0.014	−0.036
	(0.262)	(0.264)
Household head finished secondary school or higher	0.346	0.350
	(0.248)	(0.261)
Number of dependents	0.091*	0.044
	(0.052)	(0.051)
Agricultural land used for production	−0.128*	−0.124*
	(0.074)	(0.075)
Number of development projects in village	0.322***	0.301***
	(0.090)	(0.088)
Number of newly built houses in village	−0.003***	−0.003***
	(0.000)	(0.001)
Distance to health centre	−0.027	−0.028
	(0.020)	(0.023)
Distance to a tarred road	−0.009	−0.012
	(0.021)	(0.023)
Distance to closest fruit company	0.114***	0.123***
	(0.040)	(0.042)
Constant	−1.268***	−1.039**
	(0.389)	(0.429)
N	323	313
Pseudo R^2	0.216	0.229

[a]Ghana cedi (₵); *significant at 10%; **significant at 5%; ***significant at 1%. Standard errors in parentheses.

Table 10.10. Probit results for standard of living indicator 2: access to electricity.

Variable	Probit	Probit with restricted sample
Worker(s) in household works for a Fair Trade certified pineapple company	0.728**	0.739**
	(0.306)	(0.300)
Income (in 1000 GHS)[a]	–0.014	–0.015
	(0.015)	(0.015)
Household head is female	–0.222	–0.347
	(0.344)	(0.358)
Age of household head	–0.019**	–0.018**
	(0.009)	(0.009)
Household head is married	0.261	0.139
	(0.318)	(0.366)
Household head finished primary school only	–0.503*	–0.544*
	(0.304)	(0.320)
Household head finished secondary school or higher	–0.297	–0.335
	(0.255)	(0.270)
Number of dependents	0.028	0.020
	(0.074)	(0.073)
Agricultural land used for production	–0.080	–0.085
	(0.071)	(0.074)
Number of development projects in village	0.158	0.166
	(0.111)	(0.108)
Number of newly built houses in village	0.002	0.002
	(0.002)	(0.002)
Distance to health centre	–0.019	–0.024
	(0.023)	(0.023)
Distance to a tarred road	–0.065***	–0.063***
	(0.025)	(0.025)
Distance to closest fruit company	–0.059	–0.050
	(0.038)	(0.036)
Constant	1.764**	1.894***
	(0.705)	(0.701)
N	321	312
Pseudo R^2	0.130	0.128

[a]Ghana cedi (¢); *significant at 10%; **significant at 5%; ***significant at 1%. Standard errors in parentheses.

Table 10.11. Selected exemplary projects financed by Fair Trade premium.

Village	Type of project	Workers in the village (from sample)	Village population
Obom	Renovation of health post and maternity ward	14	2000
Maampehia	Classroom construction for kindergarten	15	380
Otaten	Borehole renovation	10	250
Akutiaku	Borehole renovation	17	1200
Papaase No. 1 & No. 2	Construction of toilet facilities	14	2500 3000
Nsuobri	Construction of IT centre	2	1200

appreciation for these projects, we can conclude that the Fair Trade premium has the potential to contribute to community welfare.

Additionally, export-oriented companies overall may have an impact on standard of living indicators through more indirect channels. Their set-up and establishment in the area may lead to infrastructure development and service provisions on the local level as a spill-over effect of government investment promotion. It may even be the case that particularly Fair Trade certified companies indeed use their bargaining power to foster these kinds of provisions for the villages surrounding their company grounds from where they source the majority of their workforce. Overall, we conclude that particularly Fair Trade-certified companies have the potential to improve local community welfare. These potentials could further be used to contribute to other standard-of-living indicators. On average, only 22% of the households in our sample have access to improved sanitation. This could certainly be an area for further improvements.

10.6 Conclusion

Large-scale horticultural and floricultural estates catering to the export market are expanding in developing countries. Such plantations are often associated with the exploitation of workers and appalling working conditions. Increased consumer awareness has led to a rise in sustainability standards, such as Fair Trade, organic or Rainforest Alliance. With a considerable market for their produce, a number of pineapple companies in Ghana have adopted the Fair Trade label. In the context of this chapter, we evaluate the implications of Fair Trade certification for a worker's household welfare and standards of living in the broader context of rural development.

Our findings confirm that Fair Trade certification has a positive effect on household income through higher horticultural wage labour income. Further, households with worker(s) employed on Fair Trade certified plantations are able to accumulate more assets. This may partly be a result of higher horticultural wage labour income, but Fair Trade certified companies also facilitate asset accumulation via different pathways, such as reduction of expenditure constraints for workers or loan provisions. Our results further show that other living standard indicators are also positively correlated with Fair Trade certification. Living standard indicators in rural Ghana do not only depend on household income, but also on infrastructure provision at the village level. Access to clean drinking water and electricity, but also to improved sanitation, health and educational services, is often determined by investments either by local government, local companies or aid agencies. Here, Fair Trade projects can certainly play an important role in catering to community-based needs, but projects are voted upon by all workers, which in our study area are spread over a large number of villages. Often, larger villages where more workers live have a better chance of being allocated projects due to their inhabitants' votes. Smaller (and often also poorer) villages usually accommodate fewer workers and therefore potentially have less of a chance in receiving a project. Such implications for Fair Trade's project allocation may have to be considered in more detail by the certification body. Based on our findings, non-Fair Trade certified companies may also consider fostering community welfare through local projects. Direct investments at the village level through social projects can potentially contribute to the development of areas affected by the set-up of large-scale agricultural production sites and therefore also to overarching development goals.

Our data set is not free of limitations. Relying on a cross-sectional dataset for our analysis, we cannot fully correct for unobservable characteristics possibly leading to selection bias. Future research should validate our findings using time-series or panel data. Furthermore, we realize that a company individually decides to become Fair Trade certified or not, whether the reason for this decision is altruistic or based on market demand. While we account for company-level characteristics in our analysis, in our descriptive statistics we see that it is mostly larger and potentially already more successful companies that become Fair Trade certified.

Based on our findings, the perception that large-scale plantation agriculture yields only minimal social benefits may need to be revised. In fact, considering differences between companies, production practices and worker treatment is crucial for drawing viable conclusions regarding the implications of changing agricultural patterns for rural development. Certification and standards have gained prominence over the past years. In the 1990s it was highly contested within the Fair Trade movement that plantations should be able to receive Fair Trade certification. Yet, this step has allowed Fair Trade to adapt to a land tenure and production system that is increasingly important, particularly for developing countries, where governments are supporting land consolidation to increase productivity. In this context, the growing demand for stricter labour regulations and standards can support the generation of benefits for households and communities involved in plantation agriculture. Further, increasing market demand for sustainably certified products and sustaining consumer responsibility is crucial in enabling these positive developments.

Notes

(1) Defined as dummy variables that take the value of one, if the household owns at least one unit of the respective asset.

(2) Propensity score matching is based on observable characteristics that influence participation and outcome and thus does not control for bias resulting from unobservable characteristics.

(3) 570 GHS (Ghana shillings, i.e. cedi (₵)) = approximately US$140 (June 2015).

(4) Here we asked the villages about all development projects and donors present in the village. In this case development projects do not only refer to those funded by the Fairtrade Premium.

References

Caliendo, M. and Kopeinig, S. (2005) Some practical guidance for the implementation of propensity score matching. Discussion paper series of the Institute for the Study of Labour. IZA, Bonn, Germany.

Cramer, C., Johnston, D., Oya, C. and Sender, J. (2014) *Fair Trade, Employment and Poverty Reduction in Ethiopia and Uganda*. Department for International Development (DFID), London.

Dragusanu, R., Giovannucci, D. and Nunn, N. (2014) The economics of fair trade. *Journal of Economic Perspectives* 28(3), 217–236.

Fairtrade International (2014) Our mission. Available at: http://www.fairtrade.net/our-vision.html (accessed 5 July 2017).

Filmer, D. and Pritchett, L. (2001) Estimating wealth effects without expenditure data – or tears: an application to educational enrolments in states of India. *Demography* 38(1), 115–132.

Fold, N. and Gough, K. (2008) From smallholders to transnationals: the impact of changing consumer preferences in the EU on Ghana's pineapple sector. *Geoforum* 39, 1687–1697.

Gatune, J., Chapman-Kodam, M., Korboe, K., Mulangu, F. and Rakotoarisoa, M. (2013) *Analysis of Trade Impacts on the Fresh Pineapple Sector in Ghana*. FAO Commodity and Trade Policy Research Working Paper No. 41. Food and Agriculture Organization of the United Nations, Rome.

Granville, B. and Telford, S. (2013) Empirical evidence from South Africa. In: Granville, B. and Dine, J. (eds) *The Processes and Practices of Fair Trade. Trust, Ethics and Governance*. Routledge, London and New York, pp. 286–340.

Harou, A., Walker, T. and Barrett, C. (2017) Is late really better than never? The farmer welfare effects of pineapple adoption in Ghana. *Agricultural Economics* 48(2), 153–164.

Kleemann, L., Abdulai, A. and Buss, M. (2014) Certification and access to export markets: adoption and return on investment of organic-certified pineapple farming in Ghana. *World Development* 64, 79–92.

Lyall, A. (2014) *Assessing the Impacts of Fairtrade on Worker-defined Forms of Empowerment on Ecuadorian Flower Plantations*. Final Report to Fairtrade International and Max Havelaar-Foundation.

Raynolds, L. (2012) Fair Trade flowers: global certification, environmental sustainability, and labor standards. *Rural Sociology* 77(4), 493–519.

Rosenbaum, P. and Rubin, D. (1983) The central role of the propensity score in observational studies for causal effects. *Biometrika* 70(1), 41–55.

Ruben, R. and van Schendel, L. (2009) The impact of Fair Trade in banana plantations in Ghana: income, ownership and livelihoods of banana workers. In: Ruben, R. (ed.) *The Impact of Fair Trade*. Wageningen Academic Publishers, Wageningen, The Netherlands, pp. 137–153.
Sekhar, C., Indrayan, A. and Gupta, S.M. (1991) Development of an index of need for health resources for Indian states using factor analysis. *International Journal of Epidemiology* 20(1), 246–250.
WHO and UNICEF (2015) *Progress on Sanitation and Drinking Water – 2015 Update and MDG Assessment*. Geneva, Switzerland.

11 Assessing the Benefits of Organic and Fair Trade Production for Small-scale Farmers in Asia

Yuhui Qiao*
College of Resources and Environmental Sciences, China Agricultural University, Beijing, China

Organic agriculture and Fair Trade have the potential to provide improved livelihood opportunities, increased income and social benefits for small-scale farmers. The combination has thus become a popular strategy to reduce poverty in many developing countries. Cases of small-scale farmers who have both organic and Fair Trade certification in Asia will be analysed and summarized in this chapter.

11.1 Combination and Complementarity of Organic and Fair Trade Production: from Theory to Practice

Small-scale farmers in developing countries of Asia, Latin America and Africa produce many bulk agricultural commodities. Most of them depend on agriculture as their main income source, but sometimes they are faced with problems of low productivity, low prices received for their products and degradation of agroecological conditions. The majority of farmers in poverty worldwide depend on this sector for their livelihood; roughly 730 million are employed in agriculture in Asia, mostly as informal workers (Setboonsarng, 2015).

11.1.1 Expected benefits of organic and Fair Trade production

'Under ethical agritrade ... the private sector is engaged [in] using market-based instruments that internalize the externalities of products and provide a sustainable solution that can incentivize small-scale farmers' (Setboonsarng, 2015). Certification schemes such as organic and Fair Trade are often proposed as a means to solve the problems of low product prices, environmentally unsustainable farming practices and lack of access to high-value markets (Halberg *et al.*, 2006a; Blackmore *et al.*, 2012). The rising demand for organic and Fair Trade products has also provided additional income-generating opportunities for farmers in poverty, particularly in the export market. Compliance with certification guidelines of production systems and certification of products is required in order to participate in such trade. In return, agri-producers receive price premiums (Harris *et al.*, 2001). In this sense, Fair Trade serves as an effective market-based development strategy that incorporates incentives through increased incomes.

In general, the conditions set forth for organic certification on the restriction/exclusion of agrochemical use in production systems impact

*E-mail: qiaoyh@cau.edu.cn

on economic and environmental sustainability. The contribution of organic certification to social benefits is not specific, since national organic standards hardly include provisions for social aspects. The social impacts of organic certification, such as education improvement due to higher income and increased social capital through organizational activities of farmers' groups, are achieved indirectly (Setboonsarng, 2015). Based on reviews of the developmental experience of organic agriculture among small-scale farmers, particularly for the export market, it appears that organic farmers have both organic and Fair Trade certifications. As Fair Trade primarily promotes social justice and accountability, it satisfactorily supplements organic certification, which deals mainly with the technical aspects of production.

11.1.2 Evidence of organic and Fair Trade production benefits in a global view

Organic agriculture, aiming to increase the sustainability of the ecosystems and human health, has developed very quickly all over the world since 2000. According to the FiBL-IFOAM publication on *The World of Organic Agriculture. Statistics and Emerging Trends 2016* (Willer and Lernoud, 2016), there were 43.7 million hectares of organic agricultural land (including in-conversion areas), which accounts for 0.9% of the total agricultural land in 172 countries worldwide. Based on the principles of health, ecology, fairness[1] and care (IFOAM, 2012), organic agriculture can, under the right conditions, contribute to environmental benefits and improve livelihoods among small-scale farmers. However, further research is needed to determine more precisely what potential benefits organic agriculture might offer in practice to small-scale farmers under different socio-economic and agroecological conditions (Halberg and Muller, 2012; IFOAM, 2012).

Evidence suggests that costs and benefits are context specific. While yield typically decreases after the initial conversion to organic farming, as the soil takes time to regenerate its own fertility, the reality for small-scale farmers may be the opposite, as there is often a yield increase when converting low-input, traditional farming systems to organic agriculture (Halberg *et al.*, 2006a). Moreover, organic farming could reduce the cash costs (payment with cash) because of the absence of purchased chemical fertilizer and pesticides. Although labour inputs may increase, farmers often experience a net gain due to the price premium they receive for selling certified organic products (Mendoza, 2004; Bakewell *et al.*, 2008).

Recently there has been a rise in dual certification of organic and Fair Trade. This helps to broaden market opportunities. Farmers converting to organic, with conversion periods, can benefit from adopting Fair Trade as synergies, e.g. reducing relative certification costs and gaining some price premium, while complying with organic standards. Thus, there are reasons to believe that the livelihoods of small-scale farmers may improve through participation in certified organic production schemes (Kilcher, 2007).

Several case studies and surveys in Asia, Latin America and Africa have shown that farmers are able to increase their income after the conversion period (Bacon, 2005; Tovar *et al.*, 2005; Bolwig *et al.*, 2009; Valkila, 2009). However, most of the studies focused on organic coffee production. Neilson (2008) studied global private regulations including value-chain restructuring and organic certification in Indonesian small-scale coffee systems. Panneerselvam *et al.* (2011) demonstrated in the cases in India that conversion to organic production improved food security and reduced indebtedness. Eyhorn *et al.* (2007) compared on-farm impacts of organic and conventional cotton production in India, but very little concrete research has been conducted to compare the costs and benefits of organic farming in Asian countries. Further empirical research is needed regarding the actual benefits for small-scale farmers based on socio-economic conditions, organizational context and degree of market access. This is the gap that our study seeks to fill.

11.2 Benefit Assessment of Organic and Fair Trade Coffee and Cotton Production for Small-scale Farmers in Asia

In 2014, Asia had 8% of the world's organic agricultural land (3.6 million hectares). China ranked first with 1.9 million hectares, followed by India (0.7 million hectares); Sri Lanka had 0.2 million hectares (0.75% of total arable land). The key organic arable crop group was cereals (mainly wheat and rice), with over 0.755 million hectares reported in total. Most cereals were grown in China and Kazakhstan. Oilseeds (mainly soybean) were also an important organic crop group grown on at least 0.443 million hectares, mainly in China and India. Almost 15% of organic farmland was used for perennial crops: mainly coconuts (almost 122,000 ha, mainly in the Philippines); coffee (113,000 ha, mainly in Indonesia); and tea (at least 58,000 ha, almost all in China) (Willer and Lernoud, 2016).

11.2.1 Coffee production in Asia

Coffee is a special agricultural commodity that has most of the sustainability certification and it was the first crop to be certified by Fair Trade and UTZ Certified. In Asia, Indonesia and Vietnam are the major coffee production countries. Indonesia exported 510,189 t (87% as Robusta)[(2)] with a value of nearly $822 million in 2009. Vietnam is the second largest exporter of green coffee after Brazil in terms of quantity and mainly produces Robusta coffee (Blackmore *et al.*, 2012). Coffee production in Vietnam grew rapidly in the mid 1980s with deforestation, which also caused severe environmental degradation.

Arabica coffee of high quality has a significant share of the value for export and also attracts more sustainabilty certification than Robusta coffee in Indonesia. Coffee with Fair Trade certification also grew rapidly in Vietnam, but there is no organic coffee being produced in Vietnam. In both countries coffee is produced by small-scale farmers (Blackmore *et al.*, 2012). Most of the small-scale farmers do not document their farming activities in their normal coffee production, but documentation is required for all the certification schemes. It is also very difficult for small-scale farmers to understand the differences among the schemes and this makes them confused. Sometimes they cannot see any significant price benefits.

Few studies have been carried out to explore the impact of different certification schemes on small-scale farmers in coffee production regions in Asia. A multiple certified cooperative in Indonesia whose members have benefited from certification could be considered as a success case (Blackmore *et al.*, 2012). The certified cooperative, with 8000 small-scale farmers, is the Koperasi Baitul Qiraat Baburraya (KBQB) in Bener Meriah, Aceh, where high-quality Arabica and speciality coffees are produced. KBQB has several certification schemes including Fair Trade, UTZ Certified, CAFÉ Practices and organic certification complying with US, Japanese and European Union (EU) standards. All of KBQB's farmers grow environmentally sustainable shade coffee and are certified organic.

With the help of USAID, a microfinance bank – the Koperasi Kredit Maju Bersama (KKMB) Savings and Loan organization – was set up for the cooperative certification. Buyers such as Starbucks and Royal Coffee offered the organic farmers the highest prices and paid 20% premiums above those for local conventional coffee to emphasize the high quality of coffee from the Aceh area, which increased new buyers from three customers in 2007 to six in 2008 (Blackmore *et al.*, 2012). This showed that certification schemes could be a tool to achieve social and economic benefits under specific conditions where circumstances are suitable for production of high quality/organic coffees (Blackmore *et al.*, 2012).

11.2.2 Cotton production and certification in India

India, the largest country of cotton production, is faced with the problems of low productivity, overuse of pesticides and high levels of indebtedness, despite government interventions to support cotton production. To solve these problems, organic cotton is a good choice and India is now the largest producer in the world, producing 195,000 t in 2010 with a 68% increase from the previous year (Textile Exchange, 2010). The Indian government tries to support cotton production as a way of increasing the income of Indian farmers and improve its benefits (Organic Exchange, 2008).

Eyhorn *et al.* (2007) did a survey of more than 30 organic and 60 conventional cotton farms in Madhya Pradesh in one Maikaal bioRe project for two years (2003 and 2004). The results showed that organic farms were more profitable than conventional farms with similar yields, and with 10–20% lower input costs (no chemicals and similar labour input) and 20% price premium. Gross margins were 52% higher in 2003 and 63% higher in 2004. Altogether, the results suggested that conversion to organic farming could improve livelihoods of smallholders while protecting natural resources (Eyhorn *et al.*, 2007). Panneerselvam *et al.* (2011) had similar results based on a study comparing organic and conventional cotton production under both irrigated and rain-fed conditions. As a result, the organic farmers had a higher net margin (premium added) under both irrigated and rain-fed condition, although it was not significant between systems. Moreover, the organic farmers incurred almost no debts, due to a low expenditure on inputs.

Blackmore *et al.* (2012) did a survey of farmers cooperating with the Zameen cotton company, which has Fair Trade and organic certification. The Fair Trade minimum support price can be a safety net: in a bad market Zameen can pay the farmers more than competitors. The organic premium is approximately 20% higher than the conventional price. Zameen organic farmers do not use chemical fertilizers and so they save 1500–3000 Indian rupees ($33–66) per acre with the shift to organic. However, the initial drop in yields in the early years after transition was over 50% (Blackmore *et al.*, 2012). All the studies mentioned that there would be income loss due to reduced yields during the conversion period. Support from stakeholders is essential for farmers to overcome obstacles in the conversion period.

11.3 Two Comparative Case Studies: Organic and Fair Trade Tea Production in China and Sri Lanka

11.3.1 Background

The major organic commodity in Asia is tea. Its production area is at least 81,640 ha, out of which organic tea was at least 58,000 ha in 2014 (Willer and Lernoud, 2016). Two cases in this study are certified organic as well as Fair Trade, which becomes increasingly common in the developing countries (Raynolds *et al.*, 2007). The rationale for choosing these two case areas (Wuyuan and Kandy) was that they constitute the largest organic tea production areas in China and Sri Lanka dominated by a large number of small-scale tea farmers (Qiao *et al.*, 2015). This allows for random sampling as well as comparison with conventional producers in neighbouring villages. Conclusions cannot necessarily be drawn from these case studies for other farming types.

11.3.2 Methodology: from survey to data analysis

China has 45% of the world's total tea-growing land area, with more than 80 million tea farmers (including workers and sales people) across the country (IFAD, 2005). By the end of 2010, there were 32,000 ha of organic tea production in China, yielding 28,000 t of organic tea and involving 650 enterprises (Shen and Wang, 2011). Wuyuan county in the southern province of Jiangxi, which has among the longest experience in certified organic tea production in China (starting in 1996), was selected as the case study area. The case study area is located in the north-eastern part of Wuyuan and has a geographical area of 122 km^2 with an average altitude of 1000 m. There are eight administrative villages, 53 natural villages and 3400 households. There are 1050 ha of arable land, including 590 ha of paddy field, 440 ha for tea production and 20 ha of dry land. The cropping system normally is rice–rapeseed in low lands, with tea, bamboo, oil tea, etc. on hillsides under a precipitation of 1600–1800 mm and average temperatures of 14.7–17.4°C.

The second case study site is in Sri Lanka, which is one of the oldest tea-producing countries in the world. In most cases the yield of these tea lands has fallen, mainly due to poor management in the 1980s. Only a few villagers have maintained their lands with proper fertilizing and pruning of tea bushes (Rote, 1986). In order to avoid increasing input costs in conventional tea production, these marginal tea lands were converted to organically grown tea in the early 1990s, owing to the influence of the organic movement helped by international non-governmental organizations (NGOs). The Kandy area is well known for its agroforestry-based home gardens, referred to as Kandyan Forest Gardens (KFGs).

These constitute the typical system of agroforestry in Sri Lanka, which has evolved with time to become almost the ideal form of land use, combining agriculture, forestry and even spontaneous livestock rearing. The Kandy area was selected for the case study in Sri Lanka.

Discussions were carried out between researchers and stakeholders to help select suitable case-study farms, using the following criteria: (i) farms that had been certified by an internationally accredited organization for more than five years; (ii) small-scale farms (those with a low asset base, operating with less than 2 ha of cropland (World Bank, 2003); and (iii) conventional farms that were geographically and economically similar to the selected organic farms, with the farm types and sizes being as similar as possible. Each case study site consisted of matched samples of organic and conventional farms. Approximately 100 households engaged in certified organic tea production were randomly selected at three levels of economic conditions. These organic farmers were engaged in cash crop production for export. In order to minimize selection bias and maximize similarities, comparable conventional farms were found in neighbouring villages with similar agroecological conditions, land use, household structure, infrastructure and distance to markets.

Trained local staff carried out the survey in 2007. The questionnaires gathered detailed socio-economic household and production data as well as qualitative data of farmers' perceptions following conversion to organic agriculture. Preliminary data analyses were presented at stakeholder verification workshops in the case areas in order to check the researchers' interpretation of data. Results were double-checked and some households were re-visited to verify certain information.

Quantitative survey data were statistically tested for differences between organic and conventional households using propensity score matching methods (Caliendo and Koepeinig, 2005) where comparisons are made by matching households with nearly identical values on a range of characteristics, such as resource endowment and educational level.

The cost and income/profit variables in Table 11.2 (see Section 11.3.4) are scaled per hectare. Variables in Table 11.3 (see Section 11.3.4) are scaled per household. Both sets of variables were considered relevant – the former (the area base) to facilitate comparisons across households, villages and farming systems, and the latter (the household level) to give realistic accounts of the significance of profit from organic crops in relation to the total household economy.

11.3.3 Socio-economic characteristics, production system and organization

Socio-economic characteristics of farms and households in each case study area are presented in Table 11.1. Household characteristics such as number of family members were similar in organic and conventional households both in Wuyuan (China) and Kandy (Sri Lanka); however, the age of the household head in 'organic households' was 3 years higher than the head of 'conventional households' in Kandy. In Wuyuan, the average age of the household head was almost the same in organic and conventional households.

Table 11.1. Socio-economic characteristics of organic and conventional households from case studies.

	Wuyuan, China		Kandy, Sri Lanka	
	Organic	Conventional	Organic	Conventional
Samples	87	57	100	57
Convert from	1996–1999	–	1998–2001	–
Family members	4.5	4.4	4.6	4.6
Household head age	52.4	51.9	54	51
Farm size (ha)	0.52	0.48	0.46	0.46
Tea area (ha)	0.16	0.18	0.46	0.46
Market orientation	Export	Domestic	Export	Domestic
Organization	Cooperative/ village structure	Independent	Cooperative/ company	Independent

In Wuyuan, the farm size within the two farming systems was similar, with an average of 0.5 ha of land and an average tea-growing area of around 0.2 ha. The main crops in this area were rice, rapeseeds, tea, tea-tree oil and bamboo. Tea was produced on hillsides while rice, rapeseeds and vegetables were planted on low or dry land. Livestock ownership was variable in all villages; most farmers had chickens and pigs for personal consumption. Typical organic households in this case had a split production system for commercial reasons: tea was produced organically on hillsides for export to the EU and the USA, while rice, rapeseeds and vegetables were produced conventionally on low lands mainly for personal food consumption. Tea from conventional farms was sold domestically. The Kandy area is well known for its agroforestry-based home gardens and normally the farm sizes are small, with an average size of about 0.5 ha. Because tea and other crops or trees are intercropped in the fields, unlike in organic farms in Wuyuan, organic households in the Kandy case study need to manage all crops organically to meet the requirement of organic regulations.

The organization of organic production and trade in Wuyuan is operated by a local private company as well as an organic tea farmers association. The company plays an important role in organic tea production, certification and trade; they select the tea farms with good surroundings/buffer zones and then contract with farmers each year. There is no cost to the farmers to become part of the organic production system and more than 500 households are involved in organic tea production, equivalent to about 20% of the farming households there. Governmental bodies also provide strong support for organic production. Currently it is difficult to sell conventional tea with a good price in China. On the other hand, organic tea fetches a premium price and organic producers can get consistent purchase orders offering stability and projectability as incentives for farmers to start organic production.

Organic agriculture in the Kandy area was initiated in the 1990s by two different pathways. The NGO Gami Seva Sevana (GSS) started credit and savings groups among poor small-scale farmers in 1994 and required that new members converted to organic agriculture and diversified their production to enhance their self-sufficiency. Farmers in this area had abandoned tea production because of low prices, but GSS encouraged organic tea cultivation. GSS kept a holistic household approach and offered a number of training courses and extension services to members. In 1993, another initiative within organic tea production had been launched in Kandy as a combination of Small Organic Farmers Association (SOFA) and a processing and trading company Bio Foods, which specialized in exporting teas purchased from small-scale farmers.

11.3.4 Economic benefits for small-scale farmers in both cases

The economic benefits of organic tea production are shown in Table 11.2. In Wuyuan, organic farms are located on high-altitude hillsides in good environments with few pests and diseases. Thus, organic farmers did not need to buy organic inputs to control pests and diseases. The main farming activities were soil fertility maintenance, digging, harvesting and weeding. Consequently, although organic tea production needed more labour inputs, the cash costs were very low ($0.77/ha) with most of the labour input provided by their own households. When the variable costs were calculated including farmers' own labour, it showed that the cost of organic tea production was about 1.5 times higher than the cost of conventional tea production ($P < 0.05$). The interviews revealed that conventional farmers applied more chemicals and less labour.

In Wuyuan, while the sale of organic tea was significantly (3.5 times) higher ($P < 0.001$) than conventional tea due to higher price premiums and market demand, organic tea also allowed farmers to fetch a significantly higher net cash income and profit ($P < 0.001$). The organic farms had positive profits per hectare; the profit of conventional farms was negative (Table 11.2) and farmers were not able to pay for their own labour input when farmers' own labour inputs were calculated in this study. For the organic households, the small tea plots (0.16 ha) provided an important income over a year. This was mentioned as an advantage by female household members during the survey(3), who preferred to work on-farm.

Table 11.2. Mean costs and profits of tea production for organic and conventional farms.

	Wuyuan, China ($/ha)		Kandy, Sri Lanka ($/ha)	
Variables (mean)	Organic farmers (*n* = 87)	Conventional farmers (*n* = 57)	Organic farmers (*n* = 100)	Conventional farmers (*n* = 57)
Total cash costs	0.77	14.49*	64.65	256.03**
Total costs including own labour	800.13**	565.26	423.26	594.20**
Total sales of tea	965.38***	277.56	1061.24	1217.40
Net cash income	964.49***	243.85	996.59	911.04
Profit from tea	74.23***	–303.97	646.00	623.19

*Significant for $P < 0.05$; **significant for $P < 0.01$; ***significant for $P < 0.001$

Table 11.3. Mean crop and household income and profit in the study cases.

	Wuyuan, China ($)		Kandy, Sri Lanka ($)	
Variables (mean)	Organic farmers (*n* = 87)	Conventional farmers (*n* = 57)	Organic farmers (*n* = 100)	Conventional farmers (*n* = 57)
Total cash costs	117.05	139.49	25.56	128.22***
Total crop value sold	239.74*	152.18	468.01	528.94
Net cash income	119.23*	11.79	442.44	400.72
Total costs including own labour	525.00	500.77	175.32	251.66***
Total crop value including own food	440.90	372.95	468.01	528.94
Profit from crops	–84.10	–127.82	292.69	277.28
Total profit from livestock	93.21	128.59	7.50	19.08
Total non-farm income	1142.31	1150.00	896.82	926.04
Total household income	1360.00	1296.54	1346.49	1329.84

*Significant for $P < 0.05$; **significant for $P < 0.01$; ***significant for $P < 0.001$

In Kandy, conventional farmers had more inputs for tea production and so the input costs were significantly (about 4 times) higher than for organic production ($P < 0.01$). Yields were relatively low in the organic tea farms due to new vegetative propagated (VP) crops: the new VP crops were still young and so not producing as much as they would when mature. In other places the yield levels between organic and conventional tea have become equal after a period of 4 years, but this cannot be verified from the data itself in Kandy. This is why the price premium obtained for organic tea of almost double the conventional tea price barely compensated for the yield difference. Thus, the net income and average profit from tea were not significantly different between organic and conventional tea production.

In Wuyuan, there was no difference in total variable cash costs (Table 11.3), but the value of crops sold and the net income for farmers was significantly higher with organic tea production than with conventional tea production ($P < 0.05$), mainly due to better prices and market access for organic tea. The profit from all crops grown by the organic tea farmers was similar to that of conventional ones; there is no significant difference in profit between the two kinds of households. The profit from all crops was negative in both organic and conventional households after subtracting the value of the farmers' own labour. This means that neither groups' income could pay for their own household's labour input, even when the value (or savings) of home-consumed food was included.

In Kandy (Table 11.3), the average value of all crops sold from the organic farms was a little lower than that from the conventional farms (the yield in organic farms is lower, but the price premium is higher than that from conventional farms), but the variable costs of cash were significantly lower for the organic farms ($P < 0.01$), resulting in a comparable level of net income for the two farming systems. Even though the costs of the farmers' own labour were higher in the organic system and the conventional farms hired more labourers from outside, the total variable costs including own labour were significantly lower in the organic farms as compared with the conventional farms ($P < 0.01$) and so there was no difference in the real profit from crops. Thus, the organic system, with a lower investment in terms of cash inputs and a higher input of the farmers' own labour, was more profitable.

In both cases, given the relatively small plots of land per household for the average tea farm, organic production in itself could not fully support an average household of five members. Table 11.3 also shows that farm income comprised only a small percentage of total household income in both farming systems: 11–15% in the Wuyuan case and 30–33% in the Kandy case. This indicates the important role of non-farm income (wage labour, remittances, and products and services not related to agricultural production) for the household. There were no differences in total household income between conventional and organic households because there was a comparable level of profits from crops overall, and also similar levels of non-farm income. Thus, there is a proven economic benefit to organic farming, though in these two case study areas it accounts for only a fraction of overall household income.

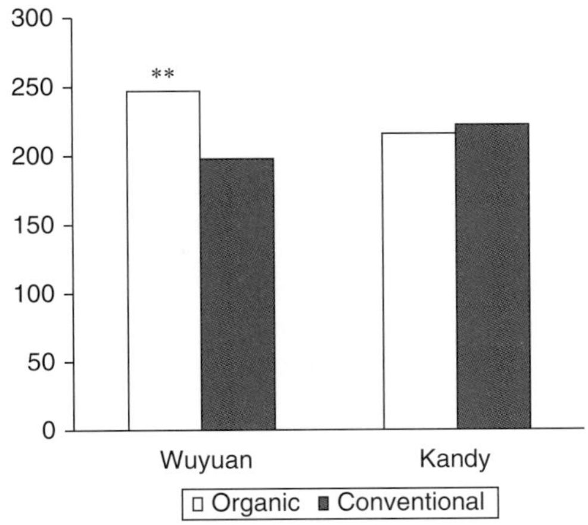

Fig. 11.1. Use of total labour per hectare in organic and conventional farms .**Significant for $p < 0.01$.

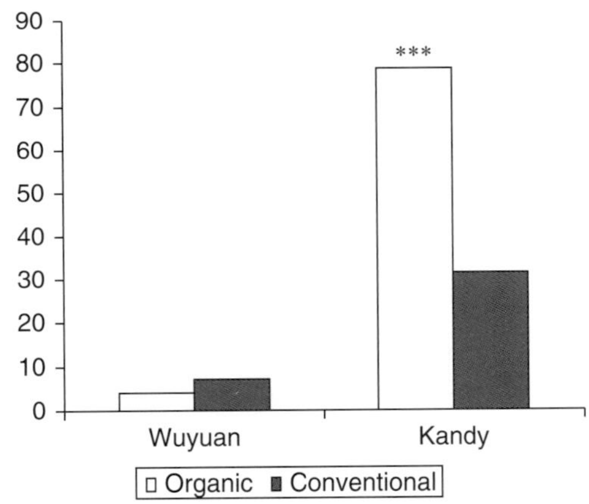

Fig. 11.2. Percentage of hired labour versus total labour used in organic and conventional farms. ***Significant for $p < 0.001$.

11.3.5 Social benefits of organic tea production

In focus group interviews in both case study sites, women noted that the increased employment opportunities in local areas owing to the organic cash-crop production was an important benefit. The survey in Wuyuan showed that the total labour input was significantly higher for organic tea production (Fig. 11.1). Women are the main labour source, and women input slightly more labour (60%) for organic tea production compared with conventional tea production (58%). For hired labour, it only accounted for 6–8% of the total labour used (Fig. 11.2); there is no difference between organic and conventional productions system and most of the labour came from their own households. In Kandy, the total labour input is almost the same in organic farms and conventional farms (Fig. 11.1), but the share of hired labour was approximately 2.5 times higher in organic than in conventional households (Fig. 11.2), resulting in a higher average use of total labour for organic farms.

In Kandy, many organic farmers benefit from jobs either at a tea factory or in the SOFA organization. The survey respondents appreciated the job opportunities for women near their residential villages that were created by the market for organic tea. In the focus group discussions, participants felt that organic farming could provide more employment opportunities for young people.

Community development is a process where community members come together to take collective action and generate solutions to common problems. Community wellbeing (economic, social, environmental and cultural) often evolves from this type of collective action being taken at a grassroots level. In 2002, after about 10 years of organic tea production, the Wuyuan case study organic farms started Fair Trade certification, as a requirement of their commercial partners in the EU. As a function of the Fair Trade scheme, from 2005 to 2008 the Wuyuan Organic Tea Farmers Association received a total of approximately $100,000 as social premiums. This amount increased annually. According to the regulations of the Fair Trade contract, the Wuyuan Tea Farmers Association used a part of the funds to improve local infrastructure such as primary processing facilities, roads, streetlights, etc. In 2006, the association set up an organic demonstration farm. Every year, the association supported the training and education of local youths from poor families in primary and middle school. Moreover, the farmers association supported the tea farmers' health care; in 2006 and 2007 the association paid for individual healthcare insurance for tea farmers and also contributed to the medical care of farmers with severe diseases such as cancer. After the earthquake in Sichuan in 2008, the Wuyuan Tea Farmers Association also donated $14,706 to the disaster area; this decision was proposed by the director of the association and approved by all the members during the meeting of the association.

In Kandy, the farmers in this study case also achieved Fair Trade certification after several years of organic tea production. While the organic price premium was paid directly to the individual farmer, the Fair Trade premium was channelled through SOFA. The total budget prepared by SOFA in relation to the Fair Trade social premium in 2006 was $50,000, which was shared among 27 village-based SOFA groups. This amounted to approximately $28 per member; however, the money could not be distributed to individual households. Most of the funds were given to community development projects proposed by a village committee and approved by the General Assembly of the association. Besides infrastructure development within SOFA itself, related to agricultural development, other areas that received support were capacity building for women of member families, and general welfare including scholarships for further education for children or farmers.

The effect of SOFA groups in terms of community development and building of social capital and organizational capacity is just as important as the amount of funds distributed. A key activity of SOFA has been to support women's self-employment and there is an increased focus on involving women in training programmes. This is linked with increased possibilities for women to gain an income from rural work either on their own land or as hired labour in organic cash crop production.

Besides the above-mentioned benefits, during the interviews in Wuyuan some tea farmers indicated that if they were not producing tea organically, some of the tea farms would have been abandoned. With organic tea, farmers can sell their product to contracted companies with a guaranteed market. Farmers in the Kandy area had neglected tea production because of low prices, but GSS encouraged organic tea cultivation and helped to identify buyers.

11.4 Conclusion

According to the above analysis, it was found that the good quality of the products, especially for coffee, in the best production regions is the premise for the sustainability certification for its additional value and higher premium. At the same time, farmers with the skills and capacity to meet the demands of group organization will be best placed to benefit from participation in certification. Farmers need to be supported during the conversion period to cover possible initial drop-offs in yields and to help them to improve organic farming practices such as composting and crop rotations. If organic markets for rotation crops can be developed, the overall profitability of farmers will improve (Blackmore *et al.*, 2012).

In order to provide evidence on the benefit of organic production, it is important to assess whether the sale of organic crops could increase farm income, while also considering any extra workload. Our tea study data showed that if we only consider costs and benefits from tea production, then the organic system had a lower investment in terms of cash inputs, as fertilizers and pesticides were not generally used. However, the organic certified farms had a higher input of households' own labour in our study. Due to the higher price premium and market access, the net income and profit is higher (in Wuyuan) or the same as the conventional system (in Kandy). The price premium of organic tea compensates for the extra labour input and/or lower yield. Elsewhere in Asia, certification studies on rice in the Philippines (Mendoza, 2004) and Bangladesh (Rasul and Gopal, 2004), tea in China (IFAD, 2005) and cotton in India (Eyhorn *et al.*, 2007) also showed that organic crops could gain higher net revenues.

The extra income from organic products was limited due to the relatively small tea plots (an average of 0.18 ha in Wuyuan and 0.46 ha in Kandy). Non-farm income was the largest portion of total household income for both organic and conventional households, so the potential economic benefit from organic farming was masked by the non-farm household income, which came mainly from household members who worked in cities. If farm income dominated the total household income, then organic agriculture would potentially significantly improve the livelihood of small-scale farmers. Together with Fair Trade schemes, this scenario would be competitive from economic, social and community development perspectives.

By increasing income and improving market access, organic farming can potentially make it more attractive for small-scale farmers to stay in their villages and work on their land (Halberg *et al.*, 2006b; Parrot *et al.*, 2006), especially where structural development would allow entrepreneurial farmers to use more land and earn a full time income, relative to conventional farming. This could be particularly beneficial for women. Organic agriculture is normally more labour intensive and this could improve the local employment of women in the companies and markets and in the processing and packaging that may take place near the areas of production. While the (small) effect of organic certification was visible in profits for individual households, the combination of organic and Fair Trade certification together resulted in livelihood improvement visible at the village level.

Perhaps organic farming is not so important from a financial perspective for a household earning off-farm income, but it also contributes to community development, especially in the case of Fair Trade–organic double certification. The extra benefit of Fair Trade certification on the community level in Wuyuan and Kandy was important; in none of the cases could organic production alone be credited for the improved livelihood and reduction in poverty that was observed in the surveyed villages.

Overall, there are clear economic benefits for small-scale farmers engaging in organic agriculture. However, the actual benefits will depend on the location and conditions of agriculture before conversion. In areas with traditional, low-input agriculture, a conversion to organic practices may benefit the farmers' livelihood and preserve landscapes with important cultural heritage more than a high-input system would. The study did not quantify the potential benefit for the farmers of reduced cash inputs. In other studies, for example from India, this has proved to be a significant benefit, because small-scale farmers often borrow money for crop inputs (fertilizer, pesticides) at very high interest rates. It was not possible to estimate the effects of the abandonment of pesticide use on farmers' health or on the environment, though it would be valuable to examine this in future studies (Panneerselvam *et al.*, 2011).

Markets are very important for the success of the different sustainability certification schemes. With such certification, the product quality will be improved with new skills and will meet certain standards and such market links could help the farmers to get support and co-investment for the production. With certain certification, price premiums will be paid by the markets

External support is also important. It could be in the form of financial support to cover the certification cost or to provide production supplies such as organic fertilizer or biopesticides. The support from various stakeholders could also include technical assistance, help with marketing and guaranteed market access.

Small-scale farmers in Asia should be organized as groups, which is essential for certification. With group organization, economies of scale could be achieved and transaction costs for developing links to markets could be decreased;

it is essential for small-scale farmers to set up effective links between themselves and their markets. Within the group organization, internal management and control systems are necessary to guarantee the practices needed to meet the requirements of certification and successful engagement with certification.

Multiple certification can be beneficial for farmers to guarantee access to different markets. It will become easier to adopt a second certification after the first one for a similar management system. Farmers converting to organic without price premium could find it advantageous to adopt Fair Trade at the same time, in order to benefit from its Fair Trade premiums while converting towards meeting organic standards. It may be essential to have organic certification as well as Fair Trade to secure market access for both.

Notes

(1) Organic agriculture should build on relationships that ensure fairness with regard to the common environment and life opportunities. Fairness is characterized by equity, respect, justice and stewardship of the shared world; both among people and in their relations to other living beings.
(2) FAO (2011) FAOSTAT trade statistics. FAO website; see http://faostat.fao.org/site/342/default.aspx
(3) The survey was carried out in July 2007 by trained local staff.

References

Bacon, C. (2005) Confronting the coffee crisis: can fair trade, organic, and specialty coffees reduce small-scale farmer vulnerability in Northern Nicaragua? *World Development* 33, 497–511.

Bakewell, S., Petra, G.L. and Charles, F. (2008) Potentials for organic agriculture to sustain livelihoods in Tanzania. *International Journal of Agricultural Sustainability* 6(1), 22–36.

Blackmore, E., Keeley, J., Pyburn, R., Mangus, E., Chen, L. and Yuhui, Q. (2012) *Pro-poor Certification: Assessing the Benefits of Sustainability Certification for Small-scale Farmers in Asia*. Natural Resource Issues, No. 25. International Institute for Environment and Development (IIED), London.

Bolwig, S., Gibbon, P. and Jones, S. (2009) The economics of smallholder organic contract farming in Tropical Africa. *World Development* 37, 1094–1104.

Caliendo, M. and Koepeinig, S. (2005) *Some Practical Guidance for the Implementation of Propensity Score Matching*. Discussion Paper no. 485, German Institute of Economic Research, Berlin.

Eyhorn, F. (2007) *Organic Farming for Sustainable Livelihoods in Developing Countries? The Case of Cotton in India*. Vdf Hochschulverlag, Zurich.

Eyhorn, F., Ramakrishnan, M. and Mader, P. (2007) The viability of cotton-based organic farming systems in India. *International Journal of Agricultural Sustainability* 5(1), 25–38.

Halberg, N. and Müller, A. (eds) (2012) *Organic Agriculture for Sustainable Livelihoods*. Earthscan/Routledge, London and New York.

Halberg, N., Knudsen, M.T., Alrøe, H.F. and Kristensen, E.S. (eds) (2006a) *Global Development of Organic Agriculture: Challenges and Prospects*. CAB International, Wallingford, UK.

Halberg, N., Sulser, T.B., Hogh-Jensen, H., Rosegrant, M.W. and Knudsen, M.T. (2006b) The impact of organic farming on food security in a regional and global perspective. In: Halbert, N., Knudsen, M.T., Alrøe, H.F. and Kristensen, E.S. (eds) *Global Development of Organic Agriculture: Challenges and Prospects*. CAB International, Wallingford, UK.

Harris, P.J.C., Browne, A.W., Barrett, H.R. and Cadoret, K. (2001) *Facilitating the Inclusion of the Resource-Poor in Organic Production and Trade: Opportunities and Constraints Posed by Certification*. A Study for the Rural Livelihoods Department, Department for International Development (DFID), UK.

IFAD (2005) *Organic Agriculture and Poverty Reduction in Asia: China and India Focus Thematic Evaluation*. IFAD Report No. 1664. International Fund for Agricultural Development, Rome.

IFOAM (2012) *The IFOAM Standard for Organic Production and Processing*. Draft version 0.2. IFOAM, Bonn, Germany. Available at: http://www.ifoam.org/about_ifoam/standards/norms/IS_V0.2_20111107formotions.doc (accessed 7 July 2017).

Kilcher, L. (2007) How organic agriculture contributes to sustainable development. In: Willer, H. and Yussefi, M. (eds) *The World of Organic Agriculture. Statistics and Emerging Trends 2007*. FIBL-IFOAM Report, 9th edn. FiBL, Frick, Switzerland; IFOAM, Bonn, Germany, pp. 82–91.

Mendoza, T.C. (2004) Evaluating the benefits of organic farming in rice agroecosystems in the Philippines. *Journal of Sustainable Agriculture* 24(2), 93–115.
Neilson, J. (2008) Global private regulation and value-chain restructuring in Indonesian smallholder coffee systems. *World Development* 36(9), 1607–1622.
Organic Exchange (2008) *Organic Cotton Farm and Fiber Report 2008*. Organic Exchange, O'Donnell, Texas.
Panneerselvam, P., Hermansen, J.E. and Halberg, N. (2011) Food security of small holding farmers: comparing organic and conventional systems in India. *Journal of Sustainable Agriculture* 35(1), 48–68.
Parrot, N., Olesen, J.E. and Høgh-Jensen, H. (2006) Certified and non-certified organic farming in the developing world. In: Halberg, N., Knudsen, M.T., Alrøe, H.F. and Kristensen, E.S. (eds) *Global Development of Organic Agriculture: Challenges and Prospects*. CAB International, Wallingford, UK, pp. 153–176.
Qiao, Y., Halberg, N., Vaheesan, S. and Scott, S. (2015) Assessing the social and economic benefits of organic and fair trade tea production for small-scale farmers in Asia: a comparative case study of China and Sri Lanka. *Renewable Agriculture and Food Systems* 31(3), 246–257.
Rasul, G. and Gopal, B.T. (2004) Sustainability of ecological and conventional agricultural systems in Bangladesh: an assessment based on environmental, economic and social perspectives. *Agricultural Systems* 79(3), 327–351.
Raynolds, L., Murray, D. and Heller, A. (2007) Regulating sustainability in the coffee sector: a comparative analysis of third-party environmental and social certification initiatives. *Agriculture and Human Values* 24, 147–163.
Rote, R. (1986) *A Taste of Bitterness. The Political Economy of Tea Plantations in Sri Lanka*. Free University Press, Amsterdam.
Setboonsarng, S. (2015) Can ethical trade certification contribute to the attainment of the Millennium Development Goals? A review of organic and fair-trade certification. In: Setboonsarng, S. and Markandya, A. (eds) *Organic Agriculture and Post-2015 Development Goals: Building on the Comparative Advantage of Poor Farmers*. Asian Development Bank, Mandaluyong City, Philippines, pp. 79–100.
Shen, X. and Wang, Q. (2011) Effect of organic certification on the economic benefit of tea enterprises based on the questionnaire of the organic tea enterprises. *Journal of Tea* 37(4), 244–249.
Textile Exchange (2010) *Farm and Fiber Report: Organic by Choice*. Textile Exchange, O'Donnell, Texas.
Tovar, L.G., Martin, L., Cruz, M.A.G. and Mutersbaugh, T. (2005) Certified organic agriculture in Mexico: market connections and certification practices in large and small producers. *Journal of Rural Studies* 21, 461–474.
Valkila, J. (2009) Fair Trade organic coffee production in Nicaragua – sustainable development or a poverty trap? *Ecological Economics* 68, 3018–3025.
Willer, H. and Lernoud, J. (eds) (2016) *The World of Organic Agriculture. Statistics and Emerging Trends 2016*. FiBL-IFOAM Report. FiBL, Frick, Switzerland; IFOAM, Bonn, Germany.
World Bank (2003) *Reaching the Rural Poor: A Renewed Strategy for Rural Development*. World Bank, Washington, DC.

12 The Impact of Certification on Material Input Costs in India

Priyanka Parvathi* and Hermann Waibel
Institute of Development and Agricultural Economics, Leibniz University Hannover, Germany

12.1 Introduction

There has been an upsurge in the growth of organic and Fair Trade markets as elaborated in Chapters 2 and 3. Also, these systems have resulted in positive impacts on rural smallholder producer livelihoods as discussed in many studies (e.g. Bacon, 2005; Kleemann and Abdulai, 2013; Parvathi and Waibel, 2015b). However, organic farming is widely criticized for low yields (de Ponti *et al.*, 2012).

Organic yields are largely reliant on quantity and quality of inputs like manure. The availability and economic accessibility of non-chemical inputs are vital for maintaining and increasing yields in organic agriculture (Brunelle *et al.*, 2015). While these non-chemical inputs are perceived to be cheaper (Seufert *et al.*, 2012), nevertheless studies have pointed out that they can also be expensive, constraining especially organic smallholder farmers from using them adequately and efficiently (e.g. Valkila, 2009; Beuchelt and Zeller, 2011). This could be a reason for low yields in organic farming.

Although a few studies have compared these costs between organic and conventional farmers (e.g. Tzouvelekas *et al.*, 2001; Pimentel *et al.*, 2005; Gündoğmuş, 2006; Valkila, 2009), they do not control for inherent selection bias in such studies. Moreover, the literature is silent on whether organic farmers also having Fair Trade certification are better able to access non-chemical inputs, given the social standards of Fair Trade (Parvathi and Waibel, 2013). In this context, we examine the effect of a double organic and Fair Trade certification on the affordability of non-chemical organic inputs among smallholder producers. We use the Indian black pepper case study elaborated in Parvathi and Waibel (2016) to investigate the certification impact on input costs. To control for endogeneity, we implement the methodology of Cater and Milon (2005) and Di Falco *et al.* (2011).

The rest of this chapter is organized as follows. Section 12.2 presents the study area, which includes a summary of the current state of black pepper production in India. Section 12.3 discusses the methodology used, followed in Section 12.4 by an examination of the results. Section 12.5 concludes the chapter with some policy recommendations.

12.2 Data

Black pepper production in India has fallen and is unable to meet the country's domestic pepper

*Corresponding author; e-mail: parvathi@ifgb.uni-hannover.de

demand. This has forced India to import pepper (Jeromi, 2007). The reasons attributed for the decline in black pepper production are primarily because of low yields due to decline in soil fertility and because of pest attacks (Gafoor *et al.*, 2007; Hema *et al.*, 2007). To improve the productivity of soil, many smallholder farmers adopted organic farming. Moreover some organic farmers also adopted Fair Trade to target international markets, as the organic market in India is still in its infancy.(1)

Data for this study were collected from Idukki district in Kerala, which produces the majority of the black pepper in India (Spices Board of India, 2009). We collected data from 100 conventional, organic and both organic and Fair Trade certified farmers (henceforth referred as double certified) in 2012.(2) As some of the farmers stored their pepper produce for future sales, we were not able to directly establish current year input expenditure incurred towards production; hence, we dropped them from our analysis. These included ten conventional, two organic and 12 double certified farmers. Therefore in our final sample we have 90 conventional, 98 organic and 89 double certified smallholders, a total of 277 farmers. We used household survey questionnaires to collect information on household characteristics, agricultural activities, off-farm employment and asset endowments, among other factors.(3)

12.3 Methodology

To control for self-selection bias we use the two-step model of Di Falco and Veronesi (2013) and Teklewold *et al.* (2013). In the first step, we use a multinomial logit model to generate selection correction terms. These generated terms are then added as regressors in the outcome regression. Secondly, we implement a counterfactual analysis model to decipher the impact of these farming systems on costs.

12.3.1 Selection and outcome equations

The farm household has the option of choosing organic or double certification as well as not being certified. If the household implements conventional farming we denote it as $s = 0$, if it follows organic or is double certified we refer to $s = 1$ and $s = 2$, respectively. Following utility theory, a household will choose a system s if it maximizes its utility over system r (Bourguignon *et al.*, 2007). We define this using a multinomial logit (MNL) model following McFadden (1973) in the first stage as described below:

$$\begin{pmatrix} \textit{probability of household } h \\ \textit{choosing farming system } s \\ \textit{over another system } r \end{pmatrix} = \frac{\exp(\beta\, sXh)}{\sum_{r=0,1}^{2} \exp(\beta\, rXh)} \qquad (12.1)$$

We add the selection correction terms generated in Eqn 12.1 as regressors in our outcome equation, which is an ordinary least square (OLS) model based on Dubin and McFadden (1984) and Bourguignon *et al.* (2007). We follow the Normalized Dubin McFadden (DMF 2) model in implementing the outcome equation as below (see equation 12.2 at the bottom of the page) where $s = 0, 1$ or 2.

Pepper cost refers to the material cost of inputs in logarithms; λ_s refers to the covariance of the errors terms in the MNL and the OLS; Ω_s represents the inverse mills ratio estimated from the probabilities in Eqn 12.1; and e_h is a random error term.

The standard errors in equation 12.2 are bootstrapped to avoid heteroskedasticity. Moreover, we follow Di Falco *et al.* (2011) and use falsification tests to identify exclusion restriction variables. As per falsification tests, we can use variables that affect the decision to choose certification in Eqn 12.1 but do not significantly impact the outcome Eqn 12.2 of conventional farmers, where $s = 0$. Moreover, despite the limitation of the Independent of Irrelevant Alternatives (IIA) limitation of the multinomial

$$\begin{aligned} \text{Pepper\_cost}_{hs} &= \text{household\_characteristics}_h \alpha_s + \text{production\_assets}_h \alpha_s \\ &+ \text{pepper\_plant\_details}_h\, \alpha_s \text{Input\_use\_dummy}_h \alpha_s + \lambda_s \Omega_s + e_{hs} \end{aligned} \qquad (12.2)$$

logit model, Bourguignon *et al.* (2007, p. 199) stated that 'selection bias correction based on the multinomial logit model can provide a fairly good correction for the outcome equation, even when the IIA hypothesis is violated'.

12.3.2 Counterfactual estimation

We draw from Carter and Milon (2005), Di Falco and Veronesi (2013) and Teklewold *et al.* (2013) to estimate counterfactual effects. In the second stage, a counterfactual analysis is implemented to determine counterfactual cost. For example, we determine the cost value of inputs of organic farmers if they were conventional and vice-versa. As we have three farming systems ($s = 0, 1, 2$), we have nine counterfactual cases as represented in Table 12.1.

We refer values in row (3) and column (3) as (3)(3), row (3) and column (2) as (3)(2) and row (3) and column (1) as (3)(1). We follow the same approach to denote values across other rows and columns in Table 12.1.

Diagonally, {(3)(3), (2)(2) and (1)(1)} represent actual cost estimations of double certified, organic and conventional farmers. We estimate these following Eqn 12.2 for $s = 2$ for (3)(3), $s = 1$ for (2)(2) and $s = 0$ for (1)(1), respectively.

We use the following Eqns 12.3a and 12.3b to estimate counterfactual cost of pepper input materials used in (3)(2) and (3)(1), respectively (see equation 12.3a and 12.3b at the bottom of the page).

Similarly we calculate counterfactual costs in (2)(3) and (2)(1) following Eqns 12.4a and 12.4b as (see equation 12.4a and 12.4b at the bottom of the page).

Following the same approach we estimate counterfactual costs in (1)(3) and (1)(2) as in Eqns 12.5a and 12.5b (see equations 12.5a and 12.5b at the bottom of the page).

Drawing from Heckman *et al.* (2001), we estimate average treatment effect on the treated (ATT) and the average treatment effect on the untreated (ATU) as shown in Table 12.1. The ATT values for double certified farmers if they opted for organic or conventional is given as the difference between (3)(3) – (3)(2) and (3)(3) – (3)(1), respectively. Likewise the ATT values for organic farmers opting to be conventional are (2)(2) – (2)(1).

$$E(\text{Pepper\_cost}_{h2} | \text{Pepper\_cost}_{h=3}) = \text{household\_characteristics}_h \alpha_2 + \text{production\_assets}_h \alpha_2 + \text{pepper\_plant\_details}_h \alpha_2 + \text{Input\_use\_dummy}_h \alpha_2 + \lambda_2 \Omega_3 \quad (12.3a)$$

$$E(\text{Pepper\_cost}_{h1} | \text{Pepper\_cost}_{h=3}) = \text{household\_characteristics}_h \alpha_1 + \text{production\_assets}_h \alpha_1 + \text{pepper\_plant\_details}_h \alpha_1 + \text{Input\_use\_dummy}_h \alpha_1 + \lambda_1 \Omega_3 \quad (12.3b)$$

$$E(\text{Pepper\_cost}_{h3} | \text{Pepper\_cost}_{h=2}) = \text{household\_characteristics}_h \alpha_3 + \text{production\_assets}_h \alpha_3 + \text{pepper\_plant\_details}_h \alpha_3 + \text{Input\_use\_dummy}_h \alpha_3 + \lambda_3 \quad (12.4a)$$

$$E(\text{Pepper\_cost}_{h1} | \text{Pepper\_cost}_{h=2}) = \text{household\_characteristics}_h \alpha_1 + \text{production\_assets}_h \alpha_1 + \text{pepper\_plant\_details}_h \alpha_1 + \text{Input\_use\_dummy}_h \alpha_1 + \lambda_1 \quad (12.4b)$$

$$E(\text{Pepper\_cost}_{h3} | \text{Pepper\_cost}_{h=1}) = \text{household\_characteristics}_h \alpha_3 + \text{production\_assets}_h \alpha_3 + \text{pepper\_plant\_details}_h \alpha_3 + \text{Input\_use\_dummy}_h \alpha_3 + \lambda_3 \quad (12.5a)$$

$$E(\text{Pepper\_cost}_{h2} | \text{Pepper\_cost}_{h=1}) = \text{household\_characteristics}_h \alpha_2 + \text{production\_assets}_h \alpha_2 + \text{pepper\_plant\_details}_h \alpha_2 + \text{Input\_use\_dummy}_h \alpha_2 + \lambda_2 \quad (12.5b)$$

Table 12.1. Counterfactual estimations (adapted from Di Falco *et al*. (2011), and Parvathi and Waibel (2015a), modified).

Farm systems	(3) OFT	(2) ORG	(1) CON	ATT	
(3) **OFT**	Actual costs for double certified farmers	Counterfactual costs estimated following (12.3a)	Counterfactual costs estimated following (12.3b)	**(a) OFT to ORG** (3)(3) – (3)(2)	**(b) OFT to CON** (3)(3) – (3)(1)
(2) **ORG**	Counterfactual costs estimated following (12.4a)	Actual costs for organic farmers	Counterfactual costs estimated following (12.4b)		**(c) ORG to CON** (2)(2) – (2)(1)
				ATU	
				(d) ORG to OFT (2)(3) – (2)(2)	
(1) **CON**	Counterfactual costs estimated following (12.5a)	Counterfactual costs estimated following (12.5b)	Actual costs for conventional farmers.	**(e) CON to OFT** (1)(3) – (1)(1)	**(f) CON to ORG** (1)(2) – (1)(1)
Heterogeneity effects	$BH_{3(A)}$ = (3)(3) – (2)(3) $BH_{3(B)}$ = (3)(3) – (1)(3) $BH_{3(C)}$ = (2)(3) – (1)(3)	$BH_{2(A)}$ = (3)(2) – (2)(2) $BH_{2(B)}$ = (3)(2) – (1)(2) $BH_{2(C)}$ = (2)(2) – (1)(2)	$BH_{1(A)}$ = (3)(1) – (2)(1) $BH_{1(B)}$ = (3)(1) – (1)(1) $BH_{1(C)}$ = (2)(1) – (1)(1)	TH (ATT – ATU) TH_1 = (a) – (d) TH_2 = (b) – (e) TH_3 = (c) – (e)	

OFT, Organic and Fair Trade double certified; ORG, Organic; CON, Conventional; ATT, Average Treatment Effect on the Treated; ATU, Average Treatment Effect on the Untreated; BH, Base Heterogeneity; TH, Transitional Heterogeneity.

Similarly, we calculate ATU values for conventional farmers if they choose to be double certified or organic as the difference between (1)(3) and (1)(1) and (1)(2) and (1)(1), respectively. The ATU for organic farmers if they opt for a double certification is given as (2)(3) – (2)(1).

We then calculate base heterogeneity (BH) following Carter and Milon (2005) and Di Falco *et al.* (2011), as depicted in Table 12.1. This is to further investigate the effect of unobservable and subjective characteristics like farming skill. We also estimate transitional heterogeneity (TH) effects to examine if the effect of certification on cost is larger or smaller for certified households as a difference between ATT and ATU, as shown in Table 12.1.

12.4 Results

Table 12.2 shows the material inputs used in the study area by conventional, organic and double certified farmers to cultivate black pepper. It needs to be noted that the input costs do not contain labour costs, as the objective of this study is to examine whether purchasing organic non-chemical inputs is expensive.

We find that the double certified farmers used significantly more manure, fertilizers and insecticide, but this could be due to the fact that there was a severe pest attack in the region during the survey year. Also organic farmers were certified organic for less than 10 years, whereas all double certified farmers were practising organic methods of cultivation for more than a decade. It appears that the double certified farmers were significantly more affected than organic farmers. This indicates that, contrary to Altieri and Nicholls (2003) and Crowder *et al.* (2010), organic farming with time requires more investment in bio-insecticides to be pest resistant and in manure to be productive. This could be because, as noted by MacFadyen *et al.* (2009) and Bengtsson *et al.* (2005), there are no significant variations between conventional and organic farms in pest control. Also, small farmers may not have the capacity to produce sufficient quantities of their own compost heaps (Branca *et al.*, 2011).

Table 12.2. Black pepper material input costs (in '000 Indian rupees) (own calculation and based on Parvathi and Waibel, 2015a).

Particulars	Conventional	Organic	Double certified
Manure/ha	3.01 ab	22.97 ac	118.76 bc
Fertilizer/ha	0.15 b	0.12 c	3.42 bc
Insecticide/ha	3.25	0.01 c	3.77 c
Fungicide/ha	0.05	0.01	0.60
Total material input cost/ha	6.46 ab	23.11 ac	126.55 bc

Values followed by a, b and c show that significant differences exist between conventional and organic, conventional and double certified, and organic and double certified, respectively. Material inputs of certified farmers refer to biofertilizers and biopesticides.

Also reduced usage of manure and fertilizers by non-adopters cannot be automatically attributed to low levels of input requirements in conventional agriculture. This could also be due to the possibility that they may not have had the means to buy adequate quantities of these inputs and hence may be rationing its application (Duflo *et al.*, 2011).

12.4.1 MNL and OLS Estimations

Table 12.3 presents the results of the MNL with a base category referring to conventional farmers. Younger but experienced farmers opt for certified systems of farming consistent with the findings of Wheeler (2008). Also owning farm tools plays a significant role in choosing to be organic or double certified. As noted by Musara *et al.* (2012) and Chouichom and Yamao (2010) a larger farm size drives farmers to venture into alternative systems of farming.

Also as anticipated, growers choose to convert to organic farming when their pepper plants are younger. As pointed out by Dadi *et al.* (2004), a shorter market distance which could result in lower transportation costs plays a vital role in certification adoption. Also if a particular farming system is perceived to be less risky, more farmers would implement it.

It is noticed that education is negatively related to double certification adoption, but this effect could be only relevant in the study region and can be attributed to the awareness campaigns conducted by the NGO in the area.

The falsification tests confirm the usage of variables distance to market and attitude towards

Table 12.3. Multinomial logit regression – selection equation (n = 277 observations).

Base Category – Conventional farmers	Organic[a]	Double certified[a]
Household characteristics		
Age	–0.064*	–0.065
	(0.035)	(0.053)
Years of schooling	–0.111	–0.386***
	(0.088)	(0.105)
Farm experience (years)	0.068***	0.047
	(0.026)	(0.043)
Total household size	–0.102	–0.334
	(0.191)	(0.211)
Dependency ratio	0.502	0.549
	(0.501)	(0.589)
Credit access (yes = 1)	–0.370	0.680
	(0.479)	(0.770)
Have off-farm income (yes = 1)	0.704	0.961*
	(0.433)	(0.498)
Production assets		
Have livestock (yes = 1)	–0.276	0.325
	(0.406)	(0.480)
Have farm tools (yes = 1)	2.104***	3.034***
	(0.447)	(0.517)
Pepper plant details		
Pepper area (ha)	1.932*	3.113***
	(1.120)	(1.187)
Age of the pepper plant (years)	–0.138***	–0.056
	(0.032)	(0.038)
Selection instruments		
Market distance (km) (log)	–0.945***	–1.111***
	(0.317)	(0.333)
Attitude towards risk	–0.808*	–0.814*
	(0.423)	(0.464)
Attitude towards soil fertility	0.796	2.009***
	(0.640)	(0.677)
Attitude towards food safety	0.632	0.386
	(0.552)	(0.651)
Constant	4.033*	4.405*
	(2.085)	(2.290)
Wald test on selection instruments (χ^2)	19.99***	24.03***
log pseudolikelihood	–195.26087	

[a]Robust standard errors clustered at panel level in parentheses. ***Significant at 1%; **significant at 5%; *significant at 10% level.

risk, soil fertility and food safety as valid exclusion restriction variables. The Wald test on these variables significantly and jointly affects the decision to adopt in MNL but does not jointly affect the input costs per hectare of conventional farmers.[(4)]

Table 12.4 presents the outcome regression. The material inputs per hectare include costs incurred on manure, fertilizer and pesticides.[(5)] For conventional farming we find that, as the age of the pepper plant increases, less expenses on chemical inputs are incurred. With respect to organic farming we notice that owning farm tools lowers input expenses but with increasing years of practising organic farming costs are likely to rise. Moreover, as expected most of the manure and pesticides purchased from markets increase expenses for all categories of farmers.

We find that almost all the selection correction terms are not significant, indicating that conventional and organic farmers will have the same input cost impacts as the double certified farmers if they adopt both these certifications. However, this is not the case for conventional farmers if they choose organic. The selection correction term is significant, suggesting that if conventional farmers choose organic, cost effects may not be the same as current organic farmers.

12.4.2 ATT, ATU and heterogeneity effects

Following Table 12.1, Table 12.5 shows the counterfactual results. As we express input costs/ha in log form, we interpret the results in percentage. The ATT results show that the input expenses of double certified farmers will increase by 81% if they opt for conventional systems. But if they de-select Fair Trade and retain organic, their costs will decline. This may be because, as the double certified farmers have been practising organic for a longer time period, they require investment in manure and biopesticides to be productive. In the case of organic farmers, ATT findings show that their costs will fall by 120% if they choose conventional farming. This also implies that biofertilizers and pesticides, if bought from outside, can be more expensive than chemical inputs.

With respect to the ATU results of organic farmers, we find that their input expenses may increase if they venture into Fair Trade certification. Contrarily, ATU results show that conventional farmers will see a decline in costs of 34%

Table 12.4. Multinomial endogenous switching regression – cost outcome equation.

Base Category – Conventional farmers	Conventional	Organic	Double certified
Dependent variable		Log of material inputs/ha	
Household characteristics			
Age	0.067	0.008	0.165
	(0.062)	(0.022)	(0.086)
Years of schooling	0.154	–0.016	–0.060
	(0.155)	(0.069)	(0.250)
Farm experience (years)	–0.039	–0.004	0.008
	(0.059)	(0.018)	(0.079)
Total household size	0.011	–0.071	0.328
	(0.295)	(0.113)	(0.406)
Dependency ratio	0.725	0.299	0.612
	(1.002)	(0.266)	(0.829)
Credit access (yes = 1)	–1. 110	0.441	0.379
	(0.980)	(0.434)	(5.093)
Have off farm income (yes = 1)	–0.534	0.343	–0.146
	(0.898)	(0.321)	(1.025)
Production assets			
Have livestock (yes = 1)	–0.429	–0.024	–1.398
	(0.821)	(0.297)	(0.960)
Have farm tools (yes = 1)	–0.802	0.656*	1.727
	(1.372)	(0.353)	(1.425)
Pepper plant details			
Pepper area (ha)	–1.669	–0.590	1.893
	(2.378)	(0.807)	(1.154)
Age of the pepper plant (in years)	–0.212**	–0.016	0.150
	(0.101)	(0.030)	(0.103)
Years practising organic farming		0.050	0.291*
		(0.046)	(0.166)
Input expenses			
Use manure (yes = 1)	8.195***	11.943***	9.597***
	(0.901)	(0.291)	(1.854)
Use fertilizer (yes = 1)	4.267	1.078	0.553
	(4.865)	(0.689)	(0.816)
Use insecticide (yes = 1)	3.715**	0.248	2.376
	(1.615)	(0.589)	(1.468)
Use fungicide (yes = 1)	–0.427	–0.830	–0.570
	(2.370)	(0.515)	(0.934)
Total labour days/ha (log)	0.030	0.165	0.492
	(0.343)	(0.326)	(0.625)
Selection bias correction terms			
_m1 (δcon)	1.774	–2.653	–2.564
	(2.683)	(1.654)	(6.254)
_m2 (δorg)	11.157**	–0.462	–4.697
	(5.626)	(0.674)	(5.210)
_m3 (δoft)	–6.950	–1.271	–0.256
	(7.129)	(1.836)	(3.043)
Constant	0.293	–4.617**	–12.281*
	(4.335)	(1.840)	(6.693)

Standard errors in parentheses. ***Significant at 1%; **significant at 5%; *significant at 10% level.

Table 12.5. Cost effects from counterfactual analysis, own calculations and based on Parvathi and Waibel (2015a).

Farm Management Regimes	(3) OFT	(2) ORG	(1) CON	ATT	
				(a) OFT to ORG	**(b) OFT to CON**
(3) **OFT**	9.379 (0.442)	8.181 (0.454)	10.194 (0.505)	1.198*** (0.223)	–0.815* (0.442)
					(c) ORG to CON
(2) **ORG**	6.549 (0.522)	5.987 (0.574)	4.780 (0.482)		1.207*** (0.308)
				ATU	
				(d) ORG to OFT	
				0.562** (0.183)	
				(e) CON to OFT	**(f) CON to ORG**
(1) **CON**	0.555 (0.575)	1.721 (0.639)	2.062 (0.494)	–1.507*** (0.324)	–0.341 (0.319)
Heterogeneity effects	$BH_{3(A)}$: 2.830*** (0.684)	$BH_{2(A)}$: 2.194** (0.731)	$BH_{1(A)}$: 5.414*** (0.699)	TH (ATT – ATU)	
	$BH_{3(B)}$: 8.824*** (0.725)	$BH_{2(B)}$: 6.460*** (0.783)	$BH_{1(B)}$: 8.132*** (0.707)	TH_1 : 0.636** (0.288)	
	$BH_{3(C)}$: 5.994*** (0.776)	$BH_{2(C)}$: 4.266*** (0.860)	$BH_{1(C)}$: 2.718*** (0.690)	TH_2 : 0.692 (0.548)	
				TH_3 : 1.548*** (0.444)	

CON, conventional; ORG, organic; OFT, organic and Fair Trade double certified. Standard errors in parentheses.
***Significant at 1%; **significant at 5%; *significant at 10% level.

and 150% if they choose organic or double certified farming, respectively.

The base heterogeneity results $BH_{3(B)}$ and $BH_{3(C)}$ as well as $BH_{2(B)}$ and $BH_{2(C)}$ show that organic farmers, if they opt for organic or double certification, will have reduced costs compared to the present category of these respective farmers. Hence, they do have some unobserved characteristics like inherent farming skill that make them cost-effective farmers irrespective of the category of farming chosen. $BH_{1(A)}$, $BH_{1(B)}$ and $BH_{1(C)}$ further reiterate that conventional farmers are more cost effective. Organic and double certified farmers would not have been able to lower costs and cultivate with the same costs as conventional farmers even if they choose conventional farming.

Transitional heterogeneity effect TH_1 is positive, demonstrating that the effects are significantly higher for the double certified farmer with respect to organic. The effects are also higher in comparison with conventional farming, although not significant (TH_2). TH_3 is positive, implying that the effect is significantly larger for organic compared with conventional farmers.

To sum up, base heterogeneity results show that conventional farmers have some unobservable characteristics that enable them to produce at lower input costs in comparison with the other two groups even under a counterfactual setting. Also, the positive transitional heterogeneity effects point out that the double certified farmers have chosen farming systems that have helped them to minimize costs in comparison with a counterfactual setting.

12.5 Conclusion

This chapter examined the impact of organic as well as a double (organic and Fair Trade) certification on material input costs. In this regard, we use household survey data of smallholder black pepper farmers collected in 2012 from India. We apply a multinomial endogenous switching regression to controls for selection bias. We also implement a counterfactual analysis including heterogeneity effects.

An important outcome is that, contrary to conventional expectations, organic and double certified systems can result in increased material input costs for smallholder farmers. However, to be cost effective the double certified farmers are better off not being conventional. The heterogeneity analysis shows that the double certified farmers benefit from cost reduction due to certification. Their costs would increase if they became conventional farmers under a counterfactual setting. This implies that perhaps certification is more essential for less competent farmers. Hence, certification can perhaps be used as a strategy to close the gap with more productive smallholder farmers.

A vital insight is that the already cost-effective farmers will benefit more from a double certification adoption. Conventional farmers who are the most cost efficient in the study region will have a further reduction in costs if they venture into double certification. This implies that organic and Fair Trade certification systems improve the competency of farmers and make them more resourceful.

Furthermore, the major lesson learnt is that, with time, organic black pepper cultivation can become vulnerable to pest attacks, leading to increased input costs. However, as pointed out by Pimentel (1993) and Pimentel *et al.* (2005), pest control under organic farming may be crop dependent. Thereby, certain crops may be more susceptible to pests under organic agriculture.

Also, it is possible that certification may result in differential gains for different categories of farmers and not be suitable for all types of crops, though organic farmers may not necessarily benefit from reduced costs if they choose double certification. Nevertheless, promoting certified farming systems as a strategy for the less productive farmers can play a critical role towards the development of sustainable agricultural systems.

Notes

(1) A detailed description of the Indian pepper scarcity is given in Parvathi and Waibel (2016).

(2) All the certified farmers received training and support from the NGO in the district which actively promoted sustainable and ethical practices of agriculture.

(3) It was observed from the data collected that the entire farm area for organic farmers were fully organic certified. There was no partial adoption.
(4) Wald falsification test results will be provided on request.
(5) The log-log model is used in comparison to other functional forms based on Akaike's information criteria (AIC) and the Bayesian information criteria (BIC) drawing from Di Falco and Chavas (2009).

References

Altieri, M.A. and Nicholls, C.I. (2003) Soil fertility management and insect pests: harmonizing soil and plant health in agroecosystems. *Soil and Tillage Research* 72(2), 203–211.

Bacon, C. (2005) Confronting the coffee crisis: can fair trade, organic, and specialty coffees reduce small-scale farmer vulnerability in northern Nicaragua? *World Development* 33(3), 497–511.

Bengtsson, J., Ahnström, J. and Weibull, A.C. (2005) The effects of organic agriculture on biodiversity and abundance: a meta-analysis. *Journal of Applied Ecology* 42(2), 261–269.

Beuchelt, T.D. and Zeller, M. (2011) Profits and poverty: certification's troubled link for Nicaragua's organic and Fairtrade coffee producers. *Ecological Economics* 70(7), 1316–1324.

Bourguignon, F., Fournier, M. and Gurgand, M. (2007) Selection bias corrections based on the multinomial logit model: Monte Carlo comparisons. *Journal of Economic Surveys* 21(1), 174–205.

Branca, G., McCarthy, N., Lipper, L. and Jolejole, M.C. (2011) *Climate-smart Agriculture: a Synthesis of Empirical Evidence of Food Security and Mitigation Benefits from Improved Cropland Management.* FAO Working paper, Mitigation of Climate Change in Agriculture Series, 3. FAO, Rome.

Brunelle, T., Dumas, P., Souty, F., Dorin, B. and Nadaud, F. (2015) Evaluating the impact of rising fertilizer prices on crop yields. *Agricultural Economics* 46(5), 653–666.

Carter, D.W. and Milon, J.W. (2005) Price knowledge in household demand for utility services. *Land Economics* 81(2), 265–283.

Chouichom, S. and Yamao, M. (2010) Comparing opinions and attitudes of organic and non-organic farmers towards organic rice farming system in northeastern Thailand. *Journal of Organic Systems* 5(1), 25–35.

Crowder, D.W., Northfield, T.D., Strand, M.R. and Snyder, W.E. (2010) Organic agriculture promotes evenness and natural pest control. *Nature* 466(7302), 109–112.

Dadi, L., Burton, M. and Ozanne, A. (2004) Duration analysis of technological adoption in Ethiopian agriculture. *Journal of Agricultural Economics* 55(3), 613–631.

de Ponti, T., Rijk, B. and van Ittersum, M.K. (2012) The crop yield gap between organic and conventional agriculture. *Agricultural Systems* 108, 1–9.

Di Falco, S. and Chavas, J.P. (2009) On crop biodiversity, risk exposure, and food security in the highlands of Ethiopia. *American Journal of Agricultural Economics* 91(3), 599–611.

Di Falco, S. and Veronesi, M. (2013) How can African agriculture adapt to climate change? A counterfactual analysis from Ethiopia. *Land Economics* 89(4), 743–766.

Di Falco, S., Veronesi, M. and Yesuf, M. (2011) Does adaptation to climate change provide food security? A micro-perspective from Ethiopia. *American Journal of Agricultural Economics* 93(3), 829–846.

Dubin, J.A. and McFadden, D.L. (1984) An econometric analysis of residential electric appliance holdings and consumption. *Econometrica: Journal of the Econometric Society* 52(2), 345–362.

Duflo, E., Kremer, M. and Robinson, J. (2011) Nudging farmers to use fertilizer: theory and experimental evidence from Kenya. *American Economic Review* 101(6), 2350–2390.

Gafoor, A., Jacob, M.J., Bhaskaran, M.K., Kokeri, S. and Nair, N.C. (2007) The Kerala state farmers' debt relief commission, Thiruvananthapuram. Recommendations made to Government of Kerala to Declare Cardamom, Coffee and Tea Distress Affected Crops, India. Available at: http://ksfdrc.kerala.gov.in/pdf/recom_idu.pdf (accessed 1 December 2014).

Gündoğmuş, E. (2006) Energy use on organic farming: A comparative analysis on organic versus conventional apricot production on small holdings in Turkey. *Energy Conversion and Management* 47(18), 3351–3359.

Heckman, J., Tobias, J.L. and Vytlacil, E. (2001) Four parameters of interest in the evaluation of social programs. *Southern Economic Journal* 211–223.

Hema, M., Kumar, R. and Singh, N.P. (2007) Volatile price and declining profitability of black pepper in India: disquieting future. *Agricultural Economics Research Review* 20(1), 61–76.

Jeromi, P.D. (2007) Farmers' indebtedness and suicides: impact of agricultural trade liberalization in Kerala. *Economic and Political Weekly* 42(31), 3241–3247.

Kleemann, L. and Abdulai, A. (2013) Organic certification, agro-ecological practices and return on investment: evidence from pineapple producers in Ghana. *Ecological Economics* 93, 330–341.
Macfadyen, S., Gibson, R., Polaszek, A., Morris, R.J., Craze, P.G., Planque, R. and Memmott, J. (2009) Do differences in food web structure between organic and conventional farms affect the ecosystem service of pest control? *Ecology Letters* 12(3), 229–238.
McFadden, D. (1973) *Conditional Logit Analysis of Qualitative Choice Behavior*. Academic Press, New York.
Musara, J.P., Chimvuramahwe, J. and Borerwe, R. (2012) Adoption and efficiency of selected conservation farming technologies in Madziva communal area, Zimbabwe: a transcendental production function approach. *Bulletin of Environment, Pharmacology and Life Sciences* 1(4), 27–38.
Parvathi, P. and Waibel, H. (2013) Fair trade and organic agriculture in developing countries: a review. *Journal of International Food and Agribusiness Marketing* 25(4), 311–323.
Parvathi, P. and Waibel, H. (2015a) Is organic agriculture and Fair Trade certification a way out of crisis? Evidence from black pepper farmers in India. *Perspektiven für die Agrar-und Ernährungswirtschaft nach der Liberalisierung*, 207.
Parvathi, P. and Waibel, H. (2015b) Adoption and impact of black pepper certification in India. *Quarterly Journal of International Agriculture* 54(2), 133–161.
Parvathi, P. and Waibel, H. (2016) Organic agriculture and fair trade: a happy marriage? A case study of certified smallholder black pepper farmers in India. *World Development* 77, 206–220.
Pimentel, D. (1993) Economics and energetics of organic and conventional farming. *Journal of Agricultural and Environmental Ethics* 6(1), 53–60.
Pimentel, D., Hepperly, P., Hanson, J., Douds, D. and Seidel, R. (2005) Environmental, energetic, and economic comparisons of organic and conventional farming systems. *BioScience* 55(7), 573–582.
Seufert, V., Ramankutty, N. and Foley, J.A. (2012) Comparing the yields of organic and conventional agriculture. *Nature* 485(7397), 229–232.
Spices Board of India (2009) *Project on pepper production in Idukki district of Kerala*. Under national horticulture mission. Spices Board Government of India Report. Available at: http://commerce.nic.in/trade/SB_Pepper_Production_In_Under_NHM_English.pdf (accessed 3 July 2016).
Teklewold, H., Kassie, M., Shiferaw, B. and Köhlin, G. (2013) Cropping system diversification, conservation tillage and modern seed adoption in Ethiopia: impacts on household income, agrochemical use and demand for labor. *Ecological Economics* 93, 85–93.
Tzouvelekas, V., Pantzios, C.J. and Fotopoulos, C. (2001) Technical efficiency of alternative farming systems: the case of Greek organic and conventional olive-growing farms. *Food Policy* 26(6), 549–569.
Valkila, J. (2009) Fair Trade organic coffee production in Nicaragua – sustainable development or a poverty trap? *Ecological Economics* 68(12), 3018–3025.
Wheeler, S.A. (2008) What influences agricultural professionals' views towards organic agriculture? *Ecological Economics* 65(1), 145–154.

13 Dovetailing Fair Trade and Organic Agro-certifications in Latin America: How the Twins Can Meet?

Ricardo Fort[1]* and Ruerd Ruben[2]
[1]*GRADE Group for the Analysis of Development, Lima, Peru;*
[2]*Wageningen Economic Research, Wageningen University and Research Centre, The Netherlands*

13.1 Introduction

This chapter provides a concise overview of the historical evolution and recent trends in Fair Trade (FT) and organic certifications for producer organizations in Latin America. Since major certified agro-commodities – starting with coffee and bananas – are produced and marketed from Latin American (and some Caribbean) countries, important lessons can be drawn with respect to the potential impact for smallholder farmers and the required modifications of fair and sustainable/organic value chains that are required in order to adapt to strongly modifying production and trade conditions.

We start by outlining an FT theory of change and identifying the additional mechanisms by which organic certification can alter or complement its effects on farmers' welfare. Particular attention is given to direct impact channels (raising incomes through guaranteed minimum prices) and indirect impact channels (better input use leading to higher crop productivity and sometimes also better quality). In addition, the effects of premium investments and changes in sustainability management practices are registered as a result from investment behaviour. The implications in terms of access to market outlets (certified FT or organic sales) and specialization of farm-level resources are identified as key mediating factors for attaining sustainable impact.

In order to assess the future feasibility of combined FT–organic certification, we outline some important effects on production systems and farm household strategies that provide incentives for such adaptation. Given the changes in rural markets for labour and capital – particularly improved access to (micro) finance and increased off-farm employment opportunities – the impact of value chain certification on improvement in net factor returns becomes more important. Moreover, public-good investments from FT premiums that are used for improved agricultural production and processing practices may require changes in land use pattern (e.g. more specialization) and thus influence the likely pay-offs of certification. These contextual factors tend to modify the agency interactions within agricultural value chains and may encourage 'beyond certification' mechanisms towards further upgrading of market networks.

The remainder of this chapter provides some recent evidence from the effects of FT and organic certification on Latin American farmers and particularly focuses on the effects of FT–organic

*Corresponding author; e-mail: rfort@grade.org.pe

combinations on sustained rural competitiveness. The main question that is addressed in this chapter is whether combined FT–organic agro-production can create the required synergies for more successful access to markets for Latin American smallholder farmers, given the changes already taking place in local agricultural market and investment conditions.

13.2 Theory of Change of Fair Trade and Organic Production

It has become common practice to outline the potential effects of specific interventions through an elaborated theory of change. This includes essentially a comprehensive description and illustration of 'how' and 'why' a desired change is expected to happen in a particular context and more specifically how farmers might respond to changes in production and trade conditions.

Fairtrade International (FLO), in its publication *Fair Trade Theory of Change* (FLO, 2015), stated that the Fair Trade vision is 'A world in which all small producers and workers can enjoy secure and sustainable livelihoods, fulfil their potential and decide on their future'. In order to achieve this vision the certification established four key interventions: economic protection policies, empowerment policies, Fairtrade Premium, and the Fairtrade mark.

In addition, FLO (2015) pointed out the mechanisms by which it is expected that the interventions should have an effect on the producers and cooperatives they work with. We have attempted to simplify that original version and highlight the points where organic production could have a potentiating effect beyond the FT mechanisms. Figure 13.1 shows the pathway of change that FT generates from three of the four interventions mentioned above (the Fairtrade mark is not taken into account, since its effects are registered among consumers and intermediaries). The graph highlights the specific mechanisms through which organic production is expected to reach producer impact.

The pathway of economic protection reduces price volatility and in addition it increases price, because the guaranteed minimum price is higher than conventional prices most of the time. These two effects combined lead to an increase in gross sales revenues. On the other hand, the FT Premium enables organizations and farmers to invest in different assets, and when these investments are in productive commons (e.g. improved processing capacity, better infrastructures, etc.), individual farm-level productivity is incremented, which also increases gross sales revenues.

The empowerment policies include better ways to interact within the farmers' organization and concern FLO guidelines on how to deal with issues like gender equality and environmental care practices. A better interaction of farmers with their organizations can lead to changes in perceptions about cooperative loyalty and probably reinforce trust and fidelity. A better relation and appropriate functioning of the farmers' organization may allow for improved business strategies and sales contracts (including reduced side sales), which in turn can lead to increases in farmers' net farm income. Guidelines related to environmental practices may also influence productivity and probably incur costs of production and thereby affect net farm income. Rules that have to do with gender equality practices or child labour will definitely affect final household welfare.

The increase in net farm income described above can lead producers to take one of the following two actions. On one hand, because households now have more income, they can spend time and resources to *diversify* their productive activities (grow other products or engage in non-farm activities). However, there is also the possibility that households prefer to *specialize* in the FT product because it generates more revenues. Independently of this decision, the increase in net household income can result in an increase in investments at home and in their assets and capital, resulting in an increase in household welfare. Besides the mechanism described above, it is important to note also that the FT premium may have an effect on assets or capital accumulation when part of it is invested in community public goods.

As mentioned before, items highlighted in Fig. 13.1 are those where complementarities of FT with organic production can be expected. Firstly, it can add to the price effect given that, most of the time, products that are FT–organic receive better prices than just FT or just organic products. Secondly, changing to organic farming

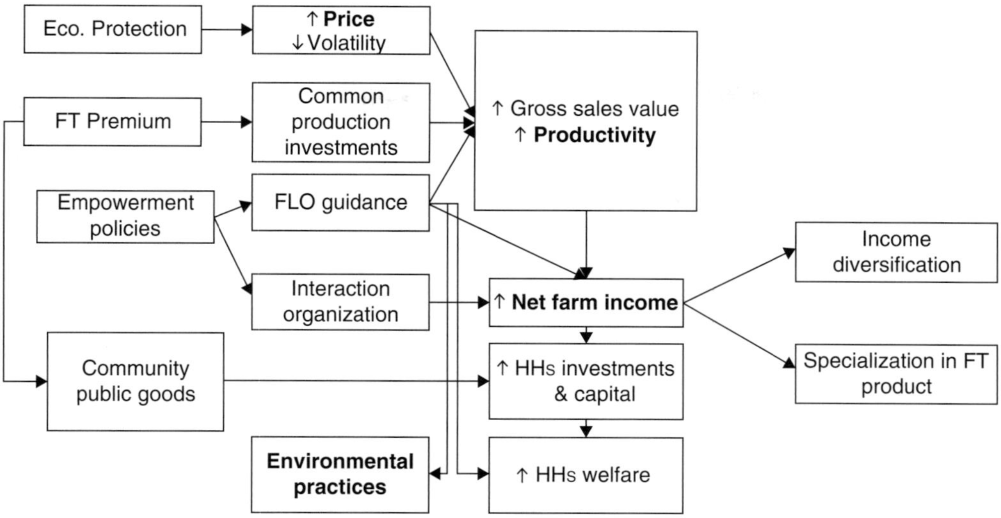

Fig. 13.1. Fair Trade and Organic certification impact pathways.

entails some costs that FT does not bear. Because of the prohibition on using several pesticides and non-organic fertilizers, usually the change to organic certification implies a reduction in productivity, at least during the initial years, and also an increase in the amount of labour needed. So, the benefit of the higher price can come at this cost, and the final effect on net farm income will depend on several factors, like the local agroecological conditions, the previous methods of production(1), the labour intensity of the crop, threats of pests and diseases, etc. Also, bearing the cost of the organic certification could be more difficult than obtaining the FT certification, as farmers usually have to wait 3 years without using prohibited products before being able to sell their produce as organic. However, it is worth noticing that FLO guidelines on environmental practices are increasingly incorporating similar rules to organic certification, so that some of the extra costs mentioned here could be present with FT certification anyway.

There is evidence in the literature that supports a large part of the primary pathway that we have identified. Probably, the increase in prices is the mechanism that has been most registered, for example by Bacon (2010), Fort and Ruben (2009a) and Wollni and Zeller (2007) in coffee, and in Fort and Ruben (2009b) for bananas. Barham *et al.* (2011) and Ruben *et al.* (2009) also found evidence of an increase in productivity thanks to FT, while only one paper by Beuchelt and Zeller (2011) found an increase in productivity thanks to organic production.

Moreover, Bacon (2010) found explicit evidence about how the rise in prices by FT reduced household vulnerability. Ruben and Zuñiga (2011) and Méndez *et al.* (2010) showed that FT increased the income of the farm and household, while Jaffee (2008) found an increase in household welfare due to the rise in net income and the cooperative investments in community public goods. Utting-Chamorro (2005) found similar evidence of women's empowerment through FT.

As to the discussion between the final specialization or diversification effects, Arnould *et al.* (2009) and Ruben and Zuñiga (2011) found that FT farmers in Latin America tended to specialize in the certificated product, while Becchetti and Costantino (2005) found that producers of FT in Kenya tended to have a greater diversification of crops compared with producers without this certification. These secondary responses tend to depend mostly on local market circumstances (relative prices and market outlets of FT/organic products versus other products) that may be modified if agricultural transitions take place. The competitive advantage of FT–organic can thus be subject to change and therefore the relative importance of the earlier outlined impact pathways could also be modified.

13.3 Recent Evidence of Fair Trade and Organic Certification Effects on Latin American Farmers

In the following sections, we provide an overview of several studies that use quantitative information and methods that include control groups or counterfactuals, for analysing the effects of FT certification or joint FT–organic certifications in Latin America. Although many of these studies examine farmers' organizations that have FT and/or organic certifications, there are just a few of them that include an evaluation design that allows for a differentiation of the FT and the organic effects in order to see how they interact.

Most of the available studies simply compare the results of farmers in these organizations with non-organized farmers without any certification, or farmers in organizations that sell only as conventional production. In the latter cases, even though there might be selection problems with the control group, we are only able to see effects of FT–organic certification as a whole.

Section 13.3.1 reviews and summarizes the main findings of these field studies, especially including reports that allow for a comparison between FT–organic and organic-only coffee cooperatives in Latin America. Section 13.4 then describes in more detail the studies by Ruben and Fort (2012) and Fort and Ruben (2009a, b) where we can analyse more carefully potential complementarities of these two certifications. We studied banana producers' organizations in the north of Peru and coffee cooperative producers in the central highlands of Peru and in both cases were able to construct an impact evaluation design that allows them to analyse complementarities of the FT certification with organic production.

13.3.1 Review of recent Latin American certification impact studies

A list of studies and initial information about their analytical approach and the results are shown in Table 13.1. The high relative importance of

Table 13.1. Summary of Latin American coffee and banana certification impact studies.

Product	Certification	Country	Effect	Study
Banana	Fair Trade	Ecuador	Positive productivity effect	Ruben *et al.* (2009)
Banana	Fair Trade	Costa Rica	Positive assets effect	Zuñiga Arias and Saenz Segura (2009)
Banana	Fair Trade and Organic	Peru	Positive price effect, positive productivity effect	Fort and Ruben (2009b)
Coffee	Fair Trade	Peru, Nicaragua, Guatemala	Positive price effect	Arnould *et al.* (2009)
Coffee	Fair Trade	Mexico	Positive price effect	Jaffee (2008)
Coffee	Fair Trade	Nicaragua	Positive price effect	Ruben and Zuñiga (2011)
Coffee	Fair Trade and Organic	Peru	No effect	Ruben and Fort (2012)
Coffee	Fair Trade and Organic	Nicaragua	Positive price effect	Bacon (2010)
Coffee	Fair Trade and Organic	Mexico	Positive price effect	Weber (2011)
Coffee	Fair Trade and Organic	Costa Rica	Positive price effect	Wollni and Zeller (2007)
Coffee	Fair Trade and Organic	Mexico, Central America	Positive price effect	Méndez *et al.* (2010)
Coffee	Fair Trade and Organic	Mexico	Positive price effect, positive productivity effect	Barham *et al.* (2011)
Coffee	Fair Trade and Organic	Nicaragua	Positive price effect, positive productivity effect (only organic)	Beuchelt and Zeller (2011)

coffee and banana FT production in Latin America has a correlate with the concentration of studies with quantitative analysis from the region. For the banana sector, Ruben *et al.* (2009) analysed the functions of production of two groups of growers in Ecuador, one with FT and one without certification. Estimating the production function of each group separately, but with the same variables, the study concluded that the FT mechanism generated higher yields, which translated directly into higher household income. Another study by Zuñiga Arias and Saenz Segura (2009) estimated the effect of having FT certification for banana plantations in Costa Rica. This study used a propensity score matching method and found an effect of FT on total household assets and on educational expenditures. In addition, an interesting finding of the study was that the FT Premium was more appreciated when handled collectively and used for investments in projects collectively agreed than when delivered individually as part of their wages.

Within the coffee sector, Arnould *et al.* (2009) analysed the effects of FT in Peru, Nicaragua and Guatemala, using a sample of small, medium and large cooperatives in each country, comparing them with unorganized coffee producers. The study showed that producers participating in FT received a higher selling price and an increase in the volume of sales, which in turn resulted in more coffee specialization. Jaffee (2008), who studied the impact of FT on two cooperatives with indigenous coffee producers in Mexico between 2001 and 2005, found similar results. FT producers received higher prices (though not enough to attract new producers), as well as realizing increased sales volume, yield and income level of the family. Jaffee also found an increase in household welfare, specifically in higher protein consumption (cheese) and better food security. Bacon's (2010) study in Nicaragua showed that being able to access the FT market reduced farmers' livelihood vulnerability because it allowed them to face crises in the conventional market.

Ruben and Zuñiga (2011) compared FT coffee producers with conventional producers and producers with two other private certification regimes (Rainforest Alliance and Café Practices) in Nicaragua, matching producers according to their structural characteristics. The study showed that FT producers had in general a higher household income (though less than Café Practices) and tended to be more specialized in coffee than producers in the other groups. An important aspect of the study was that it showed that FT had an important effect on the strengthening of organizations, as indicated by the pathway in the previous section.

However, this was not reflected in greater fidelity to the system (unlike other certifications), reinforcing the side sales problem encountered by Ruben and Fort (2012) in the Peruvian coffee cooperatives.

Weber (2011) analysed the effect on the price of FT–organic production in Mexico using a matching scheme with conventional producers. The study found that the combined FT–organic certification had a significant effect on the price, but that membership cost to be part of the organization and the household-level costs for certification made it an unattractive scheme for small producers. Wollni and Zeller (2007) conducted a study to see the effect on the price when selling to specialty markets (which means the adoption of quality standards, including FT and organic) using cooperatives in Costa Rica. As in the previous case, it was not possible to measure the marginal effect of having FT over organic, and the results showed again an increase in prices, which demonstrates the importance of economic protection interventions.

Mendez *et al.* (2010) analysed the effect of organic, FT and FT–organic coffee in Guatemala, El Salvador, Nicaragua and Mexico, surveying 469 producers in 18 cooperatives. Their results showed an increase in prices compared with conventional producers in the three cases analysed and an increase in the access to credit (to grow coffee) in the FT and FT–organic producers. Barham *et al.* (2011) in Mexico revealed that the producers of FT–organic coffee received a higher price than conventional producers (though this difference was marginal) and had higher yields that generated higher gross income from the sales of coffee.

Finally, Beuchelt and Zeller (2011) collected data for all cooperatives in northern Nicaragua in 2007 and then classified them according to their type of certification and market channel into conventional, organic and FT–organic certified cooperatives. Certified cooperatives have at least 5 years under certification. From that list, they randomly selected half of the cooperatives in each market channel, which resulted in seven cooperatives, with one of them having conventional as well as organic producers. In total, 327 cooperative members were randomly selected and classified into three groups according to their production: conventional, organic and FT–organic. A household-level questionnaire was applied to all of them.

FT–organic coffee was found to achieve the highest farm-gate prices, followed by organic coffee in comparison with conventional prices. Organic production processes required fewer purchased inputs but were far more labour intensive. Due to constrained availability of family labour, additional wage labour had to be hired, which offset saved input costs. The higher prices of certified coffees compensated for the production costs but failed to increase per hectare gross margins and profits in the case of FT–organic farmers compared with conventional producers. However, organic producers had higher yields and thus experienced an increase in per hectare gross margins and profits. Due to smaller coffee areas and large family size, the increase in gross margins did not result in improved per capita net coffee incomes for organic certified producers. Also, FT–organic certified producers did not have higher per capita net coffee incomes than conventional producers (Beuchelt and Zeller, 2011).

Beuchelt and Zeller (2011) showed that higher farm-gate prices did not lead necessarily to higher per capita net coffee income, as yield levels, production costs, family size and land size, as well as labour availability played important mediating roles. Despite the not-so-clear results regarding the profitability of the organic certified production system, the study showed that the price mechanism was enhanced by interaction of FT–organic and it also claimed that productivity does not necessarily decrease the use of organic production when it occurs in a context of low initial technology.

13.4 Detailed Case Studies on Complementarities between FT and Organic Certifications

This section presents in more detail two studies where we are able to register the complementarities between FT and organic certifications. These studies differ from many of the previous ones presented in the way that the control group was selected. The registered outcomes are therefore considered robust.

13.4.1 Banana producers' organizations in northern Peru

Banana production for export in Peru started in the northern region of Piura at the beginning of the 1990s with the arrival of a few international traders to the region. First attempts to deliver the product to US and European markets failed because of quality problems and bad coordination between the different agents in the chain. Given the highly appropriate conditions of the Chira Valley to develop organic production, the Agricultural Ministry started in 1999 a programme to promote the production and export chain for organic banana from the region. The programme promoted producers' associations and provided them with technical assistance as well as credit for the purchase of organic fertilizers and packing equipment.

As a result, several trading enterprises started to work in the valley from the year 2000, increasing the volume of organic banana exports by more than 30 times its initial level in less than 5 years. At the time of the study, there were only three export firms operating in the valley: (i) Biocosta, which started in 2002 and provided products for T.Port in Germany; (ii) the Dole subsidiary Copdeban in Peru, which started in 2001 and provided organic bananas to the US market; and (iii) Grupo Hualtaco, which started in 2002 (previously called Biorganika) and concentrated more than 80% of their sales in organic FT bananas to the European market.

At the same time, there were around 14 organizations of producers selling bananas for export to these traders. Most of them were associated with one of two networks in the valley: (i) the Central Piurana de Bananeros Organicos (CEPIBO), functioning since 2004 with five member organizations; and (ii) the Red de Pequeños Productores de Banano Organico Comercio Justo (REPEBAN), which started in 2005 with six member organizations. One of the valley's oldest producer associations – which does not belong to any of these networks – is the Asociación de Productores de Banano Orgánico del

Valle del Chira (APVCH), established in 2001 by a group of farmers who are members of the Peasant Community of Querecotillo and Salitral. The Community has existed since 1820 and is located in the Department of Piura, Province of Sullana, Districts of Querecotillo and Salitral. It currently counts more than 6000 peasants as affiliated members. Land is formally owned by the Community as a whole (title officially registered) but it is distributed equally among its members, each receiving around 1.5 ha.

APVCH has a strong relationship with the Grupo Hualtaco with whom they have worked since its formalization, and the latter assisted them in obtaining their organic and FT certifications. APVCH has 241 associates; 95% of them have been organic certified since 2002 and the rest have obtained certification since 2006. FT sales also started in 2002 and have been steadily increasing since then. This producers' association was selected as the treatment group to evaluate the impact of FT, given the relatively longer period that their members had been able to benefit from FT engagement.

We decided to select two different control groups for this study. Given the idiosyncrasy of farmers belonging to the Peasant Community, and their structural difference with other farmers in the region who acquired land through the Agrarian Reform, both control groups had to be composed of Community farmers. The first group consisted of farmers participating since 2003 in the Asociacion de Productores de Banano Organico de Salitral (APBOS), located in the District of Salitral. This association had 173 members and obtained FT certification at the beginning of 2007 but had not yet made any sales under this trademark when data was collected. Therefore, the first control group consisted of farmers who belonged to the same community, were part of a producer's organization, and had already obtained FT certification but were not using it yet. In that sense, we tried to reduce as much as possible all selection biases that might affect output indicators.

The second control group was composed of banana producers belonging to the Community, but that were not associated in any organization (consequently, they did not have FT) and did not have organic production of banana. This category of farmers can be considered as conventional banana producers. The main difference with the other groups was that these farmers were not integrated to the export chain and sold only to local intermediaries.

We collected information for 50 APVCH producers, 110 APBOS producers (Control-1) and 40 non-associated producers of conventional banana (Control-2), for a total of 200 surveys. For the selection of APVCH and APBOS farmers, we made use of the complete list of members of these organizations, containing information on total farm size and the size of their banana plantation. Sample selection for Control-1 was done in two steps. First we randomly selected farmers from the APVCH group. Thereafter, we restricted the universe of farmers from APBOS to the ones with total farm size and size of their banana plantations within the range of the APVCH sample, and then we made a random sampling from this group.

For the selection of non-associated conventional farmers (Control-2) we implemented another strategy. After finishing a questionnaire with a farmer from APVCH, the surveyor asked him to recommend two neighbouring farmers who were non-organic banana producers, but did not belong to any association and sold their product to local intermediaries. The surveyors then visited each of them and used some 'filter' questions before starting the interview. These initial questions provided confirmation on the required farmer characteristics (i.e. community member, non-organic banana producer, did not belong to any association, sale to local intermediaries). Field supervisors checked the questionnaires every night to ensure that farmers from Control-2 indeed matched the required characteristics.

The estimation method selected for the study was propensity score matching (PSM) estimation. After presenting the main characteristics and outcome variables for all groups, and comparing their means, we estimated the propensity score by using a probit model. Only exogenous variables that influence the participation decision but are not influenced by FT participation should be included in this model. The propensity score was estimated for each farmer in the complete sample by using the regression's predicted probability of having FT certification (see Appendix 13.1 for details). With the distribution of the propensity scores we proceeded to identify the regions of 'common support'. The matching estimation is performed only for observations on this common support. Matching estimators contrast the outcome of a treated individual with outcomes of comparison group members, and differ not only in the way the neighbourhood for each treated individual is defined but also with respect to the weights assigned to these neighbours. We used three different matching algorithms to check the robustness of the results to the method applied: kernel matching, one-to-one matching, and nearest neighbour matching.

In terms of the comparison between FT and organic banana producers from APVCH and non-FT organic producers from APBOS, we observe a significant difference in net household income which is mainly driven by a higher productivity in banana production. It is interesting to note that the difference in profits is obtained despite the similarity of banana prices for farmers in both organizations, as the prices for organic banana and organic–FT banana are very close.

Although we do not find significant differences in total household expenditures, FT farmers present a higher value of household assets and also receive more credit than farmers in the Control-1 group. The value of household assets acquired since 2000 is significantly higher than the one for these controls, which may imply that FT farmers invested a large proportion of their extra earnings on these improvements. Another impact of FT in terms of household wealth can be derived from the perception of higher land values. The price that an average FT household is willing to pay for an additional parcel with the same characteristics as his own plot (i.e. willingness to buy) is almost double the price for an average farmer from APBOS. In other areas, FT producers are also more identified with their organization (higher Identification index) and more convinced of their good bargaining capacity (Force index). Finally, FT banana farmers are less risk averse compared with APBOS producers, indicating their willingness to engage in higher-return activities.

Broadly similar results are obtained when comparing FT–organic producers with conventional farmers, but this time the higher income in banana production is mainly driven by the price difference rather than banana productivity. Moreover, FT farmers make substantially more land-attached investments. Conventional farmers are more affected by local market conditions during the high season when excess supply drives banana prices down. Despite this price difference, it is very important to notice that the introduction of the FT market for banana producers in the valley seems to have had an important effect on local farm-gate prices for conventional banana. As many key informants in the valley suggested, local retailers who buy conventional banana for markets in the region and in Lima appear to have been forced to increase prices in order to ensure that their providers continue to work with them.

Total household expenditures are also significantly higher for FT farmers when compared with the conventional group, reflecting an overall welfare effect of FT. Important new differences emerge in terms of the larger number of labour days (*jornales*) used for banana production, and the increased use of organic fertilizers while reducing reliance on chemical fertilizer. FT farmers not only use more family labour for this activity but also employ more wage labour from outside the household. This fact, together with the increased willingness to bear risk, seems to indicate that FT farmers are concentrating their effort on banana production instead of pursuing more income diversification activities that could reduce their risk exposure (but also decrease their expected income). Other effects related to better access to credit, higher risk acceptance and stronger engagement with local farmers' organizations are equally confirmed.

Although the original study did not report the comparison between producers with organic production and conventional producers, it is possible to calculate this with the same database and thus see a more explicit impact of the combination of FT and organic. The results using a one-to-one matching procedure are shown in Table 13.2.

As we anticipated in the discussion of impact pathways and complementarities between both types of certifications, we observe here that organic banana producers register a significantly lower productivity than their conventional neighbour producers but obtain a higher price, although only during the high season. However, this price advantage is not enough to attain higher income from banana production.

Moreover, as we consider in the comparison between FT–organic and organic-only producers,

Table 13.2. Comparison between Fair Trade and organic banana certification.

	FT–organic vs organic			FT–organic vs conventional			Organic vs conventional		
	Difference	SE	T-stat	Difference	SE	T-stat	Difference	SE	T-stat
INCOME									
Salary income	1,094	1,335	0.82	603	1,835	0.33	366	1,484	0.25
Non-salary income	−1,463	996	−1.47	−379	2,669	−0.14	13,711	12,442	1.10
Gross income, banana	4,015	1,626	2.47**	4,944	2,596	1.90*	778	2,279	0.34
Gross income, other crops	−128	518	−0.25	91	404	0.22	−52	462	−0.11
Total gross income	3,518	2,146	1.64*	5,259	3,358	1.57	14,802	11,536	1.28
Total net income	3,928	1,890	2.08**	5,573	3,426	1.63*	14,480	12,023	1.20
PRODUCTION									
Profit banana production	4,440	1,444	3.08***	5,355	2,929	1.83*	393	1,955	0.20
Banana production (kg)	8,339	3,670	2.27**	7,045	5,830	1.21	−855	5,502	−0.16
Banana productivity (kg/ha)	4,615	2,961	1.56*	3,343	5,124	0.65	−9,041	4,127	−2.19**
PRICE									
Price banana high season	0.01	0.01	0.51	0.14	0.03	5.55***	0.09	0.01	6.92***
Price banana low season	0.01	0.01	0.60	0.03	0.01	1.96*	0.00	0.01	0.22

T-stat > 2.66 = ***significant at 99%; T-stat > 2.00 = **significant at 95%; T-stat > 1.67 = *significant at 90%.

the marginal FT effect does not appear in terms of a higher price but as an increase in banana productivity, which leads to a higher profit. As discussed before, this productivity effect can be attributed to the use of the FT premium for better agricultural inputs and technical assistance as well as better infrastructure that positively affects production. The comparative results with standardized differences are show in Fig. 13.2.

13.4.2 Coffee cooperatives in Central Peru

This study evaluates the impact of FT on coffee producers in the Selva Central of Peru, in two provinces of the department of Junín (Fort and Ruben, 2009b; Ruben and Fort, 2012). Since the second half of the 19th century, this area has been one of the biggest hacienda economies in the country, with coffee plantations covering vast amounts of land. With the process of land reform in the late 1960s these provinces started to develop a cooperative model for the commercialization of coffee. As farms were still managed individually, the agrarian structure of the region could be characterized by the predominance of small and medium-size land holdings. While many of these cooperatives were dissolved during the 1990s (the so-called 'Parcellation' process), some of them managed to survive the structural reforms of that time, and some others have been reactivated recently. At the time of the study (2009), there were around ten farmers' organizations active in the provinces of Chanchamayo and Satipo in the department of Junín.

For the treatment group, we selected three cooperatives that had been FT certified for at least

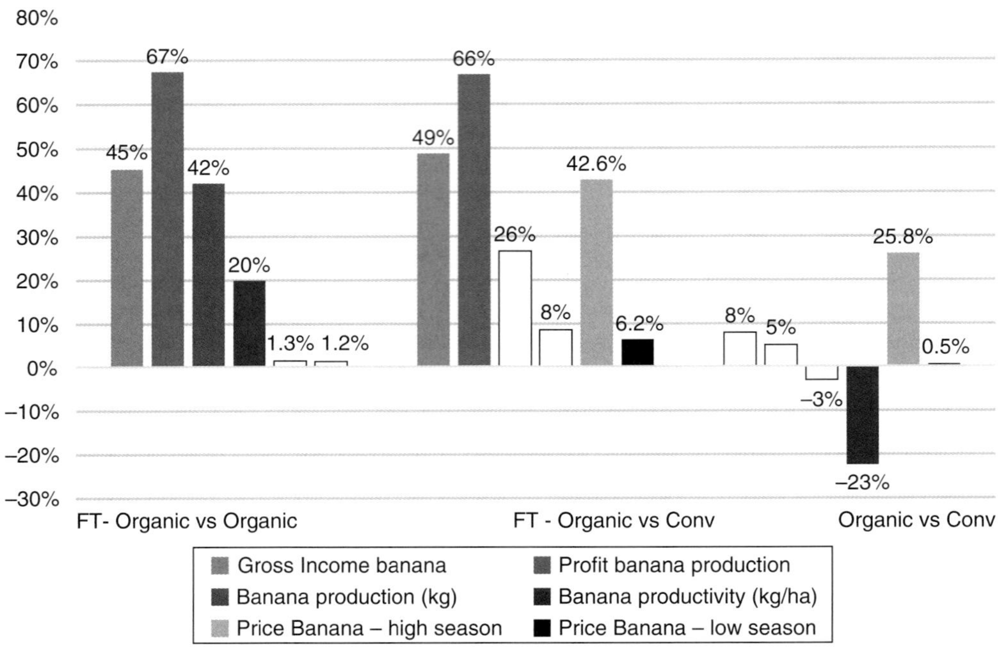

Fig. 13.2. Effects of Fair Trade versus Organic certification in banana production (Peru) (white bar represents no significant results).

3 years and were active in using the certification for their sales of coffee. An important characteristic to take into account was that all of them had at least half of their associates under organic certified coffee production and every year an important part of their production was sold under this label.

Based on the characteristics of the FT organizations under study and relying on personal interviews with their representatives, we decided to select members of three other organizations as the control group. Most of them were younger organizations and had just recently started organic production and sales under this label. What is more important is that we tried to select cooperatives that could be a good counterfactual for the FT ones in terms of the reasons for not having this certification. Hence, we looked for cooperatives where their representatives mentioned an interest in getting FT certification and the process of application had already been started. In fact, one of them already had FT certification a year before the study started, but their FT sales were very small and they had not been able to spend the FT premium yet because of a judicial problem. The other two had just started their FT application process at the time of the survey.

We could count with databases containing information on total farm sizes, size of coffee plantations, and year of organic certification (or transition) for all members of these six organizations. Using this basic information, we initially selected 60 coffee producers of each of these six organizations, 30 organic and 30 conventional, for a total sample of 360 surveys. This stratification was very important for creating better comparison groups between the FT and non-FT cooperatives and also allowed us to analyse the potential additional impact of FT when combined with the organic label (FT–organic versus organic). Sampling based on farm sizes was done firstly for members of the FT organizations, and after that we selected producers on the same area range for the control groups.

We decided firstly to separate coffee farmers in our sample, by the type of production mode that they undertook, into organic and conventional producers(2). Due to reclassification after the field survey, survey formats with incomplete information on important variables and others that were considered as outliers and as such removed from the analysis[3], the original balance between treatment and control observations in both groups was slightly modified. Table 13.3 shows some basic characteristics and pre-treatment variables for both groups of farmers and its comparison between treatment and controls.

Fair Trade farmers in both groups had on average an older and less educated head of the household, had been living for longer in their localities, had parcels further away from the district's capital, and were participating in more organizations than producers in the control group. Most importantly, land holdings of FT farmers were smaller on average than the ones for farmers in the control group, in particular for organic producers. These variables, as well as other variables that might affect the expected outcomes from FT or influence the probability of getting the FT certification, had to be taken into account in order to construct a good counterfactual for measuring the impact on FT producers. The matching estimation for getting unbiased groups of FT and non-FT farm-households can only rely on characteristics that, as such, are not influenced by participation in FT.

In order to balance the sample of FT farmers with the sample of the controls, we estimate the probability of having FT certification for the organic and conventional groups separately, based on a set of exogenous characteristics and pre-treatment variables. The probability model for organic farmers reveals a significant effect in terms of the years of residence in the locality, the number of organizations in which members of the households were participating before 2000, and the size of their coffee plantation. Only this latter effect is maintained for conventional farmers, which also show a significant difference regarding the head of household's age.

Table 13.3. Characteristics of coffee producers by group.

	Organic					Conventional				
	Fair Trade		Control		T-test	Fair Trade		Control		T-test
	N	Mean	*N*	Mean	(*P* > I t I)	*N*	Mean	*N*	Mean	(*P* > I t I)
Household characteristics										
Age head of household	91	47.3	63	41.9	0.01**	75	48.6	91	43.6	0.01**
Education head of household	91	6.4	63	7.6	0.03**	75	6.4	91	6	0.48
Family size	91	4.6	63	4.7	0.72	75	4.6	91	4.8	0.57
Residence	91	24.3	63	13.7	0.00***	75	22.1	91	16.8	0.01**
Land										
Area coffee (ha)	91	5.6	61	7.8	0.00***	74	5.7	90	6.4	0.22
Area other crops (ha)	91	0.3	63	0.5	0.48	75	0.4	91	0.3	0.70
Total productive area (ha)	91	5.9	63	8.1	0.00***	75	6.0	91	6.7	0.26
Accessibility										
Time from parcel to capital	91	103.8	60	131.7	0.00***	74	109.5	90	122.1	0.25
Time from house to parcel	91	24.5	61	23.7	0.91	74	21.0	90	20.6	0.94
Time from house to capital	90	98.7	62	108.7	0.40	75	101.1	91	114.7	0.26
Initial wealth										
Value household assets until 1999	91	321	63	159	0.56	75	295	91	208	0.73
Value agricultural assets until 1999	91	658	63	1268	0.39	75	288	91	891	0.17
Participation										
Number of organizations before 2000	91	1.3	63	0.6	0.00***	75	1.2	91	1.0	0.04**

*Significant at 10%; **significant at 5%; ***significant at 1%.

Based on the results of these regressions we proceed to estimate the propensity score for each group (organic and conventional) and identify the regions of 'common support'. Three different matching algorithms were used to analyse the stability of the results: kernel matching, one-to-one matching and nearest three neighbour matching. Given the small size of our sample (in particular of the sample in the control groups) and the strong restrictions placed on the common support, we use the results from the kernel matching to discuss our findings.

Table 13.4 presents the principal results of the PSM estimation. For both groups of farmers (organic and conventional) the results do not show any significant effect of FT involvement in terms of higher household income. Productivity levels for FT–organic farmers seem to be slightly higher than their counterparts but no significant difference could be found, though a negative and significant difference was obtained for FT conventional farmers.

The lack of a real price difference between FT and non-FT producers in both groups seems to be the main limitation for observing higher benefits. The still reduced market for FT sales in the region and the high local prices for conventional and organic coffee paid during the year of the survey as a consequence of a production shortage can at least in part explain this fact. Apparently, FT prices are increasingly considered as a regional floor price offered by local traders to all coffee farmers and thus non-FT farmers reap similar benefits as part of an externality effect.

Even though there is no significant difference in household expenditures for FT producers in either the organic or conventional groups, both of them present higher levels of animal stocks and an increment in their value of agricultural assets in the last years. In terms of their general wellbeing, farmers in older FT cooperatives appear to be better off than the ones in cooperatives with a recent FT involvement. Additionally, FT farmers in the conventional group have also managed to invest more in house improvements and land-attached infrastructure than their counterparts. The improvements made in relation to (organic) coffee production might reveal an effect of FT in terms of providing a more stable income to farmers that enables a gradual shift towards more specialized (organic) farming. However, the maintenance of higher animal stocks is revealing a still present reluctance to full specialization and the preference for a more diversified portfolio.

Finally, the lack of many expected effects from FT can at least partially be attributed to the deficient distribution and use of the FT premium as perceived by our sample of farmers. The fact that only 23% of the total number of interviewed FT producers claim to get any benefit from the use of this premium is a clear indication. Cooperatives might have to find new ways for a more effective distribution of the premium between their members so that its use can help to boost productivity levels and allow a sustainable increase of their welfare.

13.5 Discussion and Outlook

In this chapter we have presented a concise overview of the farm household impact as registered

Table 13.4. Principal coffee impacts (Fort and Ruben, 2009b).

	FT–organic versus organic		FT–conventional versus conventional	
Variable	Difference	Significance	Difference	Significance
Production gross coffee	698		–1467	
Productivity gross coffee	394		–327	**
Price average	0.20		0.05	
Household expenditures	–1395		138	
Value of animals stock	290	*	669	**
Value agricultural assets	2001	**	1412	
Made house improvement	–0.08		0.15	**

in several quantitative impact studies that compared certified farmers (FT or organic; FT–organic) with conventional (non-certified) farmers. We also included some more detailed assessments of the possible complementary effects of FT–organic certification in coffee and banana production in Peru, relying on disaggregated field data regarding production, yields, input use, prices and net profits. This enabled us to make a more balanced assessment of the likely impact pathways that generate changes in farmers' livelihoods, as well as the secondary effects that may occur due to adjustments in farming systems and marketing arrangements.

Firstly, looking at the different impact pathways (as outlined in Section 13.2), it appears that the 'protection mechanism' that allows farmers to obtain better prices for their products seems to register positive effects on farmers' income in almost all cases (Swinnen, 2007). However, it is important to notice that this result might be driven by methodological flaws. Using as a control group conventional farmers that may have different structural characteristics (in terms of farm size, education, coffee tree density) and are sometimes not even part of an association, may create a selection bias in favour of certified farmers that might have been better off even without certifications.

In more specific terms, the effects of FT certification through other pathways of yields, productivity and net income are more clearly acknowledged for banana producers than in the coffee sector. This difference is probably related to a more direct use of the FT premium in the banana case (purchase of improved inputs and technical assistance that lead to higher productivity), whereas in coffee cooperatives the premium is frequently invested in social services or maintained as a reserve fund. A tendency of farmers to specialize more household resources in the certified product appears in both cases, thus increasing their vulnerability to market volatility.

Secondly, we analysed the potential effects of combined FT–organic certification. This combination usually helps farmers to obtain a higher price when compared with only-FT or only-organic products. However, the evidence of the synergy for the combination for generating outcomes in terms of obtaining higher productivity and finally also higher net income is still less conclusive. Moreover, the Peruvian case study for coffee cooperatives shows that the price effect may not always hold when certifications are combined.

The Peru banana case study provides a nice example where an appropriate use of the FT premium for enhancing FT practices can offset the reduction of (land and labour) productivity when applying organic cultivation techniques. That mechanism – together with a higher price – resulted in higher income. However, this effect does not always occur. The literature study on coffee producers in Nicaragua shows that higher farm-gate prices did not necessarily lead to higher per capita net coffee income, as changes in yield levels, production costs and labour use and land use might occur simultaneously (Beuchelt and Zeller, 2011). Despite the not-so-clear results regarding the profitability of the organically certified coffee production system, results show that especially the price mechanism is enhanced by the interaction of FT–organic coffee, and also that productivity does not necessarily decrease using organic production when it occurs in a context of low initial technology. But due to smaller cultivated coffee areas and larger family size, the improved margins do not translate into higher per capita income.

Thirdly, certification also incites some secondary behavioural effects arising from certification that are increasingly registered. Several published papers refer to rather limited welfare effects from certification that tend to be related to limited 'loyalty' of farmers to sell their coffee to the cooperative in cases when conventional prices are close to the certified ones and cash in hand is an important issue. The certified cooperatives are frequently not able to make direct payments to farmers upon delivery of their coffee. Moreover, pre-finance (for buying inputs and paying wage labour) is only scarcely provided as part of the certification package. It is therefore relevant to compare FT–organic certification with other alternative systems that provide finance to cooperatives and enable both quality upgrading as well as increased loyalty between value chain partners.

Pre-finance can be considered as an alternative strategy for improving yields, quality and income. Several studies evaluate the effects of such alternative 'impact investments' on coffee cooperatives that have FT and organic certifications. A recent study in the Cajamarca region demonstrated that finance is critical for loyalty (or fidelity) of producers (Fort *et al.*, 2016). Thanks to the availability of impact capital, cooperatives can implement a policy for offering an advance payment of up to 80% of the estimated value of the coffee deliveries, and the remaining 20% in later months after obtaining the end price from buyers. This mechanism enables greater specialization by certified coffee producers and increases the commercial link that the producers maintain with the cooperative, which in turn generates improvements in total household incomes.

In summary, we may conclude that the original impact pathways for FT–organic certification (through prices, premium, empowerment or environment) show highly varied outcomes. Price effects can be downsized by higher input costs or stagnating yields, whereas farmers who shift towards organic production face higher labour costs, at least initially. More importantly, all certification pathways tend to disregard the importance of pre-finance, both for enabling the transition to organically certified farming, as well as for increasing loyalty with the cooperative organization.

We might even consider the current growing appreciation for impact investment as an important counterfactual for certification (ANDE, 2016). This type of funding opportunity is still scarce in the region, but it has the potential to improve access to short-term (production) credit and longer-term (investment) finance for smallholder farmers and their organizations that cannot bear the cost of becoming certified and that need operating funds for enabling their members to improve quality and productivity. Moreover, such financial schemes can accompany better the transition towards organic production, not only covering the initial payment for obtaining the certification, but also recurrent costs during subsequent transition years when productivity is still low and before obtaining the higher certified price.

These 'beyond certification' alternatives become important to support the required agricultural and rural transformations taking place in Latin America (Berdegué *et al.*, 2015). The growth of middle-class producers engaged in markets for growing cities presents new and attractive local outlets opportunities for quality products that are not necessarily certified. At the same time, better infrastructure and lower transaction costs tend to reduce the importance of certification, particularly for small-scale and poorer farmers. The entrance of wholesale enterprises and supermarkets, and their contract-farming and out-grower schemes, creates new market opportunities for smallholder farmers and support for farmers organizations that currently engage in FT and organic certifications, offering them interesting alternative opportunities for improving their wellbeing through modified market exchange.

Notes

(1) There are several examples of poor farmers that were able to make the change to organic agriculture very smoothly as they never used pesticides or other prohibited inputs before ('organic by default'), but they needed to implement substantial local substitute inputs (from waste, manure or crop residues) in order to compensate for nutrient losses.

(2) Even though this differentiation was considered from the beginning of the sampling procedure, with the use of the information in the *Padrones de productores* (producers list) the classification had to be readjusted given that some farmers had coffee parcels under both production systems. In those cases, we classified organic farmers as the ones with at least half of their total coffee area under this type of production.

(3) Most observations considered as outliers were farmers with productivity levels completely out of the possible range, and within the control group some farmers that reported selling some coffee with the FT label outside their cooperative.

References

ANDE (2016) *The Impact Investing Landscape in Latin America*. Latin American Private Equity & Venture Capital Association (LAVCA), and LGT Impact Ventures, New York.

Arnould, E., Plastina, A. and Ball, D. (2009) Does fair trade deliver on its core value proposition? Effects on income, educational attainment and health in three countries. *American Marketing Association* 28(2), 186–201. doi: 10.1509/jppm.28.2.186.

Bacon, C.M. (2010) A spot of coffee in crisis: Nicaraguan smallholder cooperatives, Fair Trade networks, and gendered empowerment. *Latin American Perspectives* 37(2), 50–71. doi: 10.1177/0094582X09356958.

Barham, B.L., Callenes, M., Gitter, S., Lewis, J. and Weber, J. (2011) Fair trade/organic coffee, rural livelihoods, and the 'agrarian question': southern Mexican coffee families in transition. *World Development* 39(1), 134–145. doi: 10.1016/j.worlddev.2010.08.005.

Becchetti, L., and Costantino, M. (2005) *The Effects of Fair Trade on Marginalised Producers: an Impact Analysis on Kenyan Farmers*. University of Vergata, Rome.

Berdegué, J., Bebbington, A. and Escobal, J. (2015) Growth, poverty and inequality in sub-national development: learning from Latin America's territories. *World Development* 73, 1–138. doi: 10.1016/j.worlddev.2014.10.015.

Beuchelt, T.D. and Zeller, M. (2011) Profits and poverty: Certification's troubled link for Nicaragua's organic and Fair Trade coffee producers. *Ecological Economics* 70(7), 1316–1324. doi: 10.1016/j.ecolecon.2011.01.005.

FLO (2015) *Fair Trade Theory of Change. Version 2.0*. Fairtrade International, Bonn.

Fort, R. and Ruben, R. (2009a) The impact of fair trade on banana producers in northern Peru. In: Ruben, R. (ed.) *The Impact of Fair Trade*. Wageningen Academic Publishers, Wageningen, The Netherlands.

Fort, R. and Ruben, R. (2009b) The impact of fair trade certification on coffee producers in Peru. In: Ruben, R. (ed.) *The Impact of Fair Trade*. Wageningen Academic Publishers, Wageningen, The Netherlands.

Fort, R., Jaramillo, M. and Vargas, R. (2016) Effects of impact investing on households participating in small and growing businesses (SBG): results from two case studies in Peru. GRADE Final Report.

Jaffee, D. (2008) *Mejor, Pero No Muy Bien Que Digamos: The Benefits and Limitations of Fair Trade for Indigenous Coffee Producer Households in Oaxaca, Mexico*. Paper presented at 3rd Fairtrade International Symposium, Montpellier, France, 14–16 May 2008.

Méndez, V.E., Bacon, C.M., Olson, M., Petchers, S., Herrador, D., Carranza, C. and Mendoza, A. (2010) Effects of Fair Trade and organic certifications on small-scale coffee farmer households in Central America and Mexico. *Renewable Agriculture and Food Systems* 25(03), 236–251. doi: 10.1017/S1742170510000268.

Ruben, R. and Fort, R. (2012) The impact of fair trade certification for coffee farmers in Peru. *World Development* 40(3), 570–582. doi: 10.1016/j.worlddev.2011.07.030.

Ruben, R. and Zuñiga, G. (2011) How standards compete: comparative impact of coffee certification schemes in Northern Nicaragua. *Supply Chain Management: An International Journal* 16(2), 98–109. doi: 10.1108/13598541111115356.

Ruben, R., Clercx, L., Cepeda, D. and de Hoop, T. (2009) Widening and deepening Fair Trade impact of banana production in El Guabo association, Ecuador: a production function analysis. In: Ruben, R. (ed.) *The Impact of Fair Trade*. Wageningen Academic Publishers, Wageningen, The Netherlands.

Swinnen, J.F.M. (2007) *Global Supply Chains, Standards and the Poor*. CAB International, Wallingford, UK.

Utting-Chamorro, K. (2005) Does Fair Trade make a difference? The case of small coffee producers in Nicaragua. *Development in Practice* 15(3–4), 584–599. doi: org/10.1080/09614520500075706.

Weber, J.G. (2011) How much more do growers receive for fair trade-organic coffee? *Food Policy* 36(5), 677–684. doi: 10.1016/j.foodpol.2011.05.007.

Wollni, M. and Zeller, M. (2007) Do farmers benefit from participating in specialty markets and cooperatives? The case of coffee marketing in Costa Rica. *Agricultural Economics* 37(2–3), 243–248. doi: 10.1111/j.1574-0862.2007.00270.x.

Zuñiga Arias, G. and Saenz Segura, F. (2009) The impact of Fair Trade in the banana production in Costa Rica. In: Ruben, R. (ed.) *The Impact of Fair Trade*. Wageningen Academic Publishers, Wageningen, The Netherlands.

Appendix 13.1

This appendix shows the probit regressions for both groups and the distribution of the propensity score before and after the pairing, for the comparison of FT–Organic versus Organic and versus Conventional coffee.

Table A13.1. Probit regression.

	FT–Organic vs Organic			FT–Organic vs Conventional		
Variables	Coeff.	SE	Sign	Coeff.	SE	Sign
Age head of household	–0.015	(0.014)		0.012	(0.0189)	
Education head of household	–0.096	(0.035)	**	–0.020	(0.049)	
Family size	0.049	(0.081)		0.153	(0.123)	
Banana area	0.085	(0.242)		0.921	(0.362)	**
Area other crops	–0.594	(0.599)		–0.755	(0.923)	
Time parcel to capital	0.040	(0.011)	***	0.049	(0.016)	**
Value agricultural assets until 1999	0.002	(0.004)		–0.007	(0.003)	**
Organizations before year 2000	–0.055	(0.146)		0.156	(0.204)	
Years residing in the locality	0.000	(0.012)		–0.030	(0.018)	
Constant	0.128	(0.880)		–1.898	(1.281)	
	Number of obs = 150 LR chi2(9) = 38.95 Prob > chi2 = 0 Pseudo R^2 = 0.2071			Number of obs = 84 LR chi2(9) = 31.43 Prob > chi2 = 0.0002 Pseudo R^2 = 0.2739		

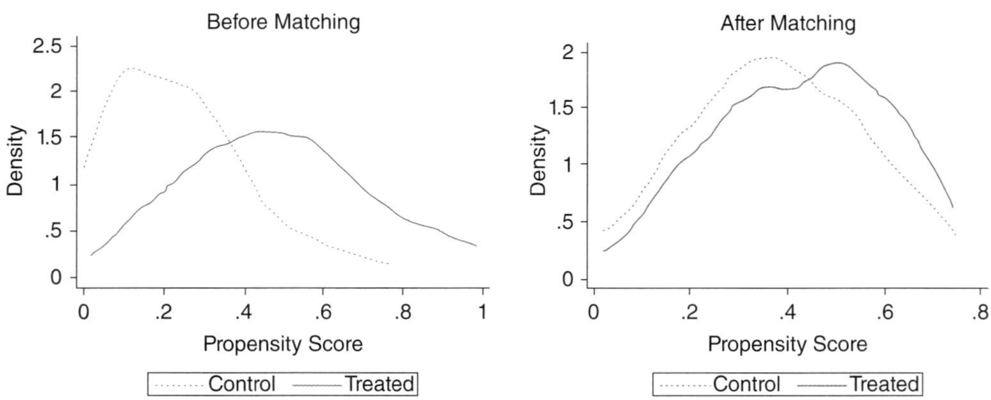

Graph A13.1: Distribution of propensity scores, FT–Organic versus Organic.

14 Certifying Coffee Cooperatives in Ethiopia, India and Nicaragua: How Far Do Small-scale Coffee Producers Benefit?

Pradyot Ranjan Jena[1,2*], Ulrike Grote[2] and Till Stellmacher[3]
[1]*School of Management, National Institute of Technology, Karnataka, India;* [2]*Institute for Environmental Economics and World Trade, Leibniz University Hannover, Germany;* [3]*Center for Development Research (ZEF), University of Bonn, Germany*

14.1 Introduction

Certification in general and coffee certification in particular are fuelled by consumer sensibility and awareness of what products they are buying and about the circumstances under which these products have been produced, processed and marketed. Increasingly consumers are willing to pay higher prices for products that meet certain required attributes. This movement has paved the way for bigger market shares of certified coffee in the major consuming countries and also created new opportunities for smallholder coffee producers in many developing countries to reap socio-economic benefits by participating in certified marketing channels.

From the data published by the International Coffee Organization (ICO), it is observed that Nicaragua, India and Ethiopia are some of the major coffee-producing countries. In 2015, Nicaragua accounted for 1.5% of global coffee production, India for 4% and Ethiopia for around 4.2%, making Ethiopia the fifth, India the sixth and Nicaragua the 12th largest producers worldwide (ICO, 2016). In Nicaragua, approximately 40% of the rural workforce depends on coffee cultivation (Valkila and Nygren, 2009). In Ethiopia, coffee is a cash crop for an estimated 15 million smallholder farmers (Tadesse *et al.*, 2013).

Fair Trade and organic certification have been increasingly promoted in Ethiopia, India and Nicaragua as a market-based instrument to reduce poverty among small-scale coffee farmers. Each of these countries has different institutional and historical dimensions of coffee production, processing and marketing, which makes comparisons among them meaningful. In Ethiopia, the current primary cooperative system was developed in the 1990s from formerly state-run Agricultural Service Cooperatives (Stellmacher, 2007; Jena *et al.*, 2012). Organized cooperatives in the coffee sector have a long presence in Nicaragua (Jena *et al.*, 2017). In contrast, organizing coffee farmers under a cooperative is a recent phenomenon in India. Such initiatives are often carried out by non-governmental organizations (NGOs) (Jena and Grote, 2017).

The major objective of this chapter is to analyse and compare the impacts of coffee certification schemes in countries such as Ethiopia, India and Nicaragua. There are four reasons to choose these three countries for analysis. Firstly, they all produce Arabica coffee(1), particularly

*Corresponding author; e-mail: pradyotjena@nitk.edu.in

dominated by small-scale production. Secondly, in all three countries a large proportion of the total output is from extensive highland forest coffee production systems. Thirdly, an increasing share of the forest coffee production of each country is subject to certification according to Fair Trade and organic standards. Fourthly, forest coffee production and certification in the three countries are undertaken within entirely different local contexts. Broadly, the product (Arabica highland (semi-)forest coffee) is overall the same in all study cases(2), as are the investigated certification standards (Fair Trade and organic for smallholders). The certified organizations (rural smallholder coffee cooperatives) are structurally similar, while the historical, political, institutional and socio-economic backgrounds are utterly different. This allows best data comparison and the evaluation of how the standardized approach of certification fits to the context specificity of smallholder coffee production worldwide. Against this background, the objective of the chapter is to investigate and compare the livelihood impacts of Fair Trade and/or organic certification for smallholder coffee farmers in primary cooperatives in Ethiopia, India and Nicaragua.

In this chapter, Section 14.2 provides a critical review of the literature related to impact evaluation of coffee certification, followed in Section 14.3 by descriptions of our case study survey areas, sample selection procedures and methodology. Findings from the quantitative analysis are given in Section 14.4, while Section 14.5 supplements the quantitative analysis with the qualitative findings. The conclusions and recommendations are furnished in the last section.

14.2 Literature Review

Several certification programmes have now entered the global mainstream coffee market. The major ones are organic, Fair Trade, UTZ, Rainforest Alliance and Smithsonian 'Bird Friendly'. Their backgrounds and objectives are, however, very different. Fair Trade, one of the most widely used standards for coffee worldwide, is defined as 'an alternative approach to conventional trade that aims to improve the livelihoods and well-being of small producers by improving their market access, strengthening their organizations, paying them a fair price with a fixed minimum, and providing continuity in trading relationships' (Giovannucci and Koekoek, 2003). Organic certifications are mainly focused on production and methods applied for processing that maintain and restore or enhance ecological sustainability, excluding the application of chemical inputs or genetically modified inputs. Bird-friendly certification mainly targets the conservation of forest cover through coffee production under the shade of a forest canopy. The main objectives of the above-mentioned certification standards differ among them. However, they all apply integrated approaches and include requirements that address both socio-economic and environmental issues.

In the past two decades among developing countries, coffee certification has become a 'flagship programme' that is the focus of research. The scale and scope of the studies in this particular sector are expanding to document how far smallholder producers actually benefit.

Several studies show that certification has a positive impact on stabilizing farm-gate producers' prices for coffee and the access to certified markets has a positive and significant effect on sale prices. Dempsey and Campbell (2006) showed for Ethiopia that Fair Trade almost doubled the value of smallholder-produced coffee, while organic certification added approximately 10% to the price received by farmers during the coffee crisis period from 2001 to 2003. Kamau *et al.* (2010) showed that households with certified cooperatives had more off-farm income, received more credit and made more capital-related investments than those in non-certified ones in Kenya. Using survey data and propensity score matching with multiple treatments, Chiputwa *et al.* (2015) showed that Fair Trade certification increased the standard of living and also reduced the depth of poverty among coffee planters in Uganda. Bacon (2005) showed that Nicaraguan farmers participating in Fair Trade and organic certification were four times better off and had less chance of losing their land during coffee price crises when compared with non-certified farmers. Pariente (2000) showed for Costa Rica that certified marketing channels, particularly Fair Trade, had the opportunity to reduce price volatility for smallholder producers when they took part in certified marketing channels. In addition to the price premium, Muradian and Pelupessy (2005) and Fromm and Dubón (2006) showed that certification was a major tool with farmers that provided better access to well-defined markets and could enable certified farmers to increase their sales. Using samples of both Fair Trade and non-Fair Trade certified coffee smallholders from Peru, Ruben and Fort (2011) found only modest direct income effect and productivity difference between certified and non-certified groups; however, they observed significant differences in input use.

There are also empirical studies with critical conclusions concerning certification. Parrilli (2001) and Tiyapongpattana (2001) in Mexico and Thailand, respectively, showed that farm-gate prices earned through certified production processes were not necessarily higher than those from conventional production approaches. Bacon (2005) showed for Nicaragua that, although Fair Trade and organic certification had a significant impact on coffee smallholders to improve their livelihood, it was not sufficient to offset other factors that reduced their general livelihood. Utting-Chamorro (2005) demonstrated limits to the extent to which Fair Trade could raise the living standard of Nicaraguan small coffee producers due to cooperatives' debt problems and volatile world market prices.

Valkila (2009) showed for Nicaragua that certified channels provided higher income for coffee smallholders compared with conventional channels but only in low-intensity coffee farming. Similarly, Beuchelt and Zeller (2011) concluded from their surveys in Nicaragua that even though farm gate prices for certified coffee were higher than those for conventional coffee, certification did not improve the livelihoods of coffee farmers. They further provided evidence that certified farmers more often hovered below the poverty line than non-certified ones. These findings have to be viewed against the background that cooperatives are smallholder organizations and Fair Trade certification particularly targets marginalized farmers.

Beyond direct impacts of certification on the livelihoods of smallholder farmers in terms of price premium, access to well defined markets and reduced price volatility, a number of indirect impacts could also be found. Several papers identified capacity building as one of the most important benefits of Fair Trade certification. Furthermore, being part of Fair Trade networks, farmers' groups could benefit from physical capital investment, technical and financial support, employment benefits or the introduction of life and medical insurances (Castro, 2001).

A study for the UK government's Department for International Development (OPM/IIED, 2000) showed that certification networks such as Fair Trade primarily depended on farmers' cooperatives structures, e.g. to monitor prices and the use of social premia and in building technical and personnel capacities(3). Hence, producers' organizations are key for the transmission and monitoring of benefits accruing from certified value chains to the individual farmers who belong to such organizations. In their work on cocoa trading in Ecuador, Nelson and Galvez (2000) identified training of the staff of the farmers' organizations in marketing skills, the strengthening of their organizational capacities and the practice of production and post-harvest techniques as the main benefits of Fair Trade certification. The findings were also supported by Hopkins (2000) and Ronchi (2002). Giovannucci and Potts (2008) showed that effectiveness and impact of certification on socio-economic and environmental sustainability of smallholder producers highly depended on the local environment in which to exhibit the proper implementation and monitoring of certification schemes.

In sum, empirical research on the direct and indirect effects of coffee certification on income and poverty is still evolving; currently, it is still characterized by individual case-based and inconclusive results. To shed more light on this, our study was designed as a combination of three case study surveys which can help us to compare our results across Ethiopia, India and Nicaragua.

14.3 Data and Methodology

14.3.1 Data

We conducted household surveys with a total of 738 coffee farmers in Jimma region in Ethiopia, Araku Valley in India and Jinotega region in Nicaragua. The selection of these survey sites was based on the fact that smallholder coffee farmer cooperatives from these regions are subjected to

Fair Trade and organic certification schemes. In all three countries, face-to-face interviews were conducted with the farming household head.

In Ethiopia, the period of primary data collection was September–November 2009. We conducted expert interviews with staff of the Oromia Coffee Cooperative Union in Addis Ababa before our field study. These interviews helped in identifying the three districts (*woreda*), namely Limmu Kossa, Gomma and Manna, within Jimma Zone of Oromia Regional State in south-western Ethiopia as the survey region. Jimma Zone is the major coffee plantation zone in Ethiopia. The Agricultural and Rural Development Office statistics of Jimma show that coffee production in the region claims about 20% of the overall export share in Ethiopia. About 30–45% of the population in Jimma are involved in coffee production, processing and marketing, directly or indirectly. As is typical in Ethiopia, Jimma is mostly dominated by smallholder coffee producers producing under traditional agroforestry systems. Coffee produced in Jimma has been branded as 'Arabica Limmu', which is an international brand of coffee attracting a premium price. Further, with the information from the Oromia Coffee Cooperative Union, six primary coffee cooperatives were selected for the study. Among the six, four of the unions had Fair Trade standards and the other two were non-certified conventional cooperatives. The research survey method followed in Ethiopia was a mix of quantitative and qualitative techniques. The data was collected through structured face-to-face interviews for the sample of 249 small coffee-producing famers among selected cooperatives. The sample consisted of 166 certified and 83 non-certified households. Additionally, expert interviews were conducted with cooperative managers and staff. Furthermore, focus group discussions (FGDs) with semi-structured guidelines were conducted to add further inputs to the data. Finally, additional expert interviews were conducted at the Ethiopia Commodity Exchange (ECX) in Addis Ababa.

In India, there are three types of coffee-growing regions: traditional areas representing the southern states of Karnataka, Kerala and Tamil Nadu; non-traditional areas consisting of Andhra Pradesh and Orissa; and the north-eastern 'seven sisters' states.(4) Among the traditional areas, Karnataka accounts for 71% of the total coffee production of India, followed by Kerala (20.5%) and Tamil Nadu (6.7%). The production share of the non-traditional areas is only 1.8%, whereas the share from the north-eastern region is negligible (CBI, 2010).

Araku district in the non-traditional coffee-growing region of northern Andhra Pradesh has been selected as a case study in India because it is one of the few places in the country where smallholder farmers who produce Arabica highland (semi-)forest coffee are organized in a coffee cooperative that is Fair Trade certified. The Araku coffee cooperative investigated in this study was founded in 2007 by an NGO to organize and assist the (mostly Adivasi tribal) smallholder coffee farmers who had been hitherto engaged in forest shifting cultivation. In 2008, the cooperative became Fair Trade certified.

A comprehensive survey was conducted in Araku district during March and April 2010 with 256 farmers purposively sampled in six villages. Four of these villages were under the coffee cooperative coverage and certified by Fair Trade standards while the other two were not certified. Of the interviewed farmers, 155 were members of the certified cooperative.

In Nicaragua, nearly 80% of the coffee is produced in the northern highlands, namely in Jinotega and Matagalpa Municipalities (IICA, 2004). Jinotega alone accounts for around 65% of the total Nicaraguan coffee production (UCA Soppexcca, 2011). In 2009, we undertook a pre-study in Managua and Jinotega in which interviews with experts from NGOs, the coffee sector and scientists were conducted. In 2010, a household survey with 233 coffee farmers was carried out in Jinotega Municipality. Approximately 8000 out of the 139,000 coffee farmers in Nicaragua are located in Jinotega Municipality. Bearing this in mind, we followed a disproportionate stratified random sample selection. In the first step of the sample selection, four cooperative unions were chosen based on the certification status of the local cooperatives operating under their jurisprudence. The first two used Fair Trade and organic certification standards, whereas the latter two were not certified and hence functioned as a control group. Thereafter, we randomly selected the primary cooperatives under these cooperative unions and finally the individual members of these primary cooperatives. Since the total number of smallholders in each primary cooperative varied, the random selection of these smallholders under each primary cooperative was disproportionate. Most of the interviewed cooperative members were coffee smallholders: 98% of them depended on coffee production as the main source of income and 69% of them possessed not more than 5 ha of land in total.

14.3.2 Methodology

Selection bias is a common econometric problem while undertaking impact evaluation analysis based on cross-sectional data (Wooldridge, 2002). Selection bias occurs because the adopters of a typical development programme (which in our study was a certification programme) sometimes self-select into the programme by virtue of certain factors that are not observed by the researcher. Such omitted variables can create distortion in estimating the impact of the development programme in question. If the selection bias is not accounted for in the estimation, it creates endogeneity in the cross-section data and thus can lead to inefficient estimations in ordinary least square (OLS), resulting in misleading interpretations. We used propensity score matching (PSM) methods for correcting sample selection bias in this study. PSM has been well documented in previous studies, such as Jena *et al.* (2012) and Jena and Grote (2017), hence we do not elaborate on the method here.

14.4 Findings for Ethiopia, India and Nicaragua

The empirical findings of this chapter are divided into three subsections. Section 14.4.1 describes the key characteristics of the smallholder coffee farmers in the three case study countries. These characteristics are grouped under productive resources that are expected to improve smallholders' income. These resources are institutional resources, human resources, physical resources and financial resources. Section 14.4.2 compares the yield, farm-gate prices obtained and net incomes generated from a unit of coffee land for the three countries. Finally, Section 14.4.3 provides empirical evidence on the impact of certification on smallholder income.

14.4.1 Key characteristics of coffee farmers

While organizing smallholder coffee farmers under the umbrella of a cooperative system is relatively common, historically rooted and politically supported in Nicaragua and Ethiopia, it is a new development in India. The key characteristics of coffee-farming smallholder households in the case study areas in Ethiopia, India and Nicaragua are described in Table 14.1. The variables are characterized as institutional, human, physical and financial resources, as well as other controls. Institutional resources refer to the respondents' certification status. This variable is used as the treatment variable in our econometric exercises. Proportionally larger samples were collected from the certified farmers population compared with the non-certified ones in all three countries. Accordingly, the percentage of certified farmers in the total sample is 70% in the Nicaraguan case, 67% in the Ethiopian and 59% in the Indian case. Among the human resource variables, average number of years of school education of the household heads is similar in all three countries, with 4.2 years in Ethiopia, 3.3 in India and 3.6 in Nicaragua. Many household heads, especially in Ethiopia and India, are illiterate. These are remarkably low figures for all three coffee-farming areas and also in intranational comparison. The low educational levels of the coffee farmers are likely to pose a constraint to the local implementation and comprehension of certification standards and procedures. The years of experience of the household head in coffee farming vary significantly among the case studies, with Ethiopian farmers having the highest average experience (20 years), followed by Nicaraguan farmers (12 years) and Indian farmers (10 years). Regarding the provision of extension services, Nicaraguan cooperatives score higher than Ethiopian and Indian cooperatives, as 92% of interviewed farmers in Nicaragua have obtained extension services compared with 43% and 41% in Ethiopia and India, respectively. This shows the differences in agricultural capacity-building efforts between the three case studies, which is instrumental in delivering the benefits of certification. With regard to labour input for coffee farming, smallholders largely make use of their own household labour force. An average of about three household members per

Table 14.1. Summary of variables.

Variable name	Description of variable	Nicaragua *(sample size: 233)*	Ethiopia *(sample size: 249)*	India *(sample size: 256)*	ANOVA test[b]	Kruskal–Wallis test[b]
Institutional resource						
Certification status[a]	1 = if household certified; 0 = otherwise	0.7 (0.5)	0.67 (0.47)	0.59 (0.49)		
Human resources						
Education	Education of head of household in years	3.6 (3.6)	4.23 (3.24)	3.34 (5)	3.78	
Experience	Years of experience of head of household in coffee farming	12.5 (9.45)	20 (13)	9.9 (9)	23.50***	
Training	Whether respondent has received any training (yes = 1; otherwise = 0)	0.92 (0.27)	0.43 (0.49)	0.41 (0.49)		1.25
Family labour	Number of family members working in coffee cultivation	3.27 (1.96)	4.2 (2.35)	2.8 (1.17)	5.32**	
Physical resources						
Size of landholding	Size of farm (ha)	5.07 (7)	1.67 (1.25)	1.85 (1)	20.67***	
Livestock	Whether household has any form of livestock (yes = 1; otherwise = 0)	0.85 (0.35)	0.88 (0.33)	0.7 (0.45)		2.36
Financial resource						
Access to non-farm income	1 if the household has non-farm income; 0 otherwise	0.43 (0.49)	0.12 (0.33)	0.84 (0.37)		15.87***
Control variables						
Age	Age of head of household in years	47.3 (13)	48 (14)	38.5 (10.3)	10.39**	
Gender	Gender of head of household (male = 1; female = 0)	0.74 (0.4)	0.9 (0.25)	0.9 (0.29)		3.78
Household size	Total number of household members	5.84 (2.4)	6.2 (1.9)	4.67 (1.65)	2.96	

Continued

Table 14.1. Continued.

Variable name	Description of variable	Nicaragua *(sample size: 233)*	Ethiopia *(sample size: 249)*	India *(sample size: 256)*	ANOVA test[b]	Kruskal–Wallis test[b]
Dependency ratio	Household members below 14 and above 65 years divided by total number of household members	0.67 (0.7)	0.69 (0.7)	0.4 (0.53)	5.79	
Distance	Distance (km) from household to selling point	21.8 (13.6)	5.2 (5.4)	2.68 (4.47)	28.35***	

[a]Nicaraguan sample consists of both organic and Fair Trade certified farmers; Ethiopian and Indian sample comprises only Fair Trade certified farmers. [b]To test the difference in statistical distribution of the variables among three countries one-way ANOVA for normally distributed variables and Kruskal–Wallis test for categorical variables have been used. F test for ANOVA and chi-squared test for Kruskal–Wallis test have been reported. ***Significant at 1% level; **significant at 5% level; *significant at 10% level.

farm are involved in coffee farming in Nicaragua and India and four members in Ethiopia. The average size of the farmland is somewhat different in each case study. Nicaraguan farmers hold an average farm size of 5 ha, Indian farmers 1.85 ha and Ethiopian 1.67 ha. Most coffee farmers in the study areas possess livestock; 87% of the Ethiopian farmers, 85% of the Nicaraguan farmers and 70% of the Indian farmers have livestock.

Non-farm income is a crucial indicator for farm income diversification and resilience and can be a cushion for reducing poverty. Our data shows that the Ethiopian coffee farms have by far the least access to such income. Only 12% reported having non-farm income. In the Nicaraguan case, 42% of the coffee farms had non-farm income. The surprise result is from India, where 84% of the coffee farms reported having non-farm income. However, since the Indian case study is undertaken in an area with a large proportion of Adivasi tribal people subject to special state programmes (such as road construction), many farm household members work as day labourers in related programmes. The daily wage is relatively low compared with other non-farm activities such as fixed-salary jobs and shop-owning. The age composition of the respondents is similar in the three case study areas with average ages between 38 and 48 years. Between 74% and 90% of the interviewees are male. In the sample, the household size differs between 6.2 people (Ethiopia), 5.8 people (Nicaragua) and 4.7 people (India). The dependency ratio is higher in Ethiopia and Nicaragua with 69% and 67%, respectively, whereas it is 40% in the Indian case. The average distance of the farm to the coffee selling point is rather large for Nicaraguan farmers with 21 km, whereas it is only 3 km on average for Indian farmers since coffee merchants tend to buy the produce directly at farm gate.

14.4.2 A comparative analysis of yields, prices and incomes

The mean values of coffee yields, prices from both cooperatives and private merchants on local markets as well as the gross margins per unit of land of coffee cultivation for the case studies in Ethiopia, India and Nicaragua are produced in Table 14.2. We compare certified and non-certified channels of marketing.

Total land area and area under coffee are largest in Nicaragua and smallest in India. While the total land area amounts to between 1.59 ha in Ethiopia to 2.8 ha in Nicaragua for non-certified or single-certified farmers (the total area for double certified farmers amounts to 4.88 ha), the area under coffee varies between 0.8 ha in India and 1.92 ha in Nicaragua. Total land area and area under coffee are generally slightly larger or about the same for Fair Trade certified

Table 14.2. Net revenues in certified and conventional channels in the case studies.

Indicators	Ethiopia		India		Nicaragua			
	Fair Trade certified	Non-certified	Fair Trade certified	Non-certified	Fair Trade certified	Organic certified	Fair Trade–organic certified	Non-certified
No. of households	166	83	155	99	79	38	46	70
Total land (ha)	1.70	1.59	1.96**	1.68**	2.13	2.84***	4.88***	2.13
	(1.20)	(1.34)	(1.19)	(0.82)	(1.65)	(2.73)	(5.20)	(1.73)
Area under coffee (ha)	1.20***	0.84***	0.90	0.82	1.60**	1.92**	1.88**	1.65**
	(0.14)	(0.09)	(0.56)*	(0.41)*	(1.85)	(1.60)	(1.26)	(1.52)
Yield (kg/ha)	871*	1,035*	530	515	1,561***	1,165	1,210	1,206
	(61.5)	(95.08)	(486)	(470.6)	(1,208)	(745.7)	(862)	(988)
Price from cooperative (local currency)	3.30	3.21	18.17[a]	n.a.	1,112	1,390***	1,428***	1,111
	(0.17)	(0.10)	(0.6)		(153)	(167)	(142)	(124)
Price from private trader (local currency)	5.54***	6.41***	93.65	90.23	987	942	941	975
	(0.21)	(0.28)	(10.18)**	(8.5)**	(142)	(120)	(143)	(178)
Net income from coffee (local currency)	4,441	5,415	10,965	7,213	19,682	22,337**	22,762**	19,079
	(460)	(600)	(11,732)***	(6970.5)***	(18,338)	(18,053)	(25,344)	(23,349)

[a]1 kg dry coffee requires 6 kg red coffee cherries. ***Significant at 1% level; **significant at 5% level; *significant at 10% level. Figures in parentheses are standard deviations.

farmers as for non-certified farmers. For Ethiopia, the average values and the mean separation test in Table 14.2 indicate that Fair Trade certified farmers in our sample have a yield rate of 871 kg/ha as compared with the yield rate of 1035 kg/ha of non-certified farmers. In the Araku district in India, the average coffee yield is found to be much lower than in Ethiopia, but it is not statistically and significantly different between the two groups. The certified farmers harvested 530 kg/ha whereas the non-certified farmers yielded 515 kg/ha. The high standard deviation in yield is partly due to the difference in individual farmers' engagement in their coffee farms and partly due to the lack of access to the necessary inputs. When the respondents were asked to state three major reasons for their low yields in coffee production, they mentioned coffee plant diseases, insufficient rain or no access to irrigation, and high labour costs. In the Jinotega Municipality in Nicaragua, average coffee yields are generally significantly higher than in Ethiopia and India. Within Nicaragua, Fair Trade certified coffee farmers have significantly higher yields than both the conventional and organic coffee farmers. While the mean harvest of Fair Trade coffee farmers is 1561 kg/ha, conventional farmers harvested only 1206/kg. Farmers who are both organic and Fair Trade certified produced a mean harvest of 1210 kg/ha and only-organic certified farmers 1165 kg/ha. The relatively low yields of organic farmers are not surprising: they complained during our open interviews that since they were not allowed to use chemical fertilizer, they solely depended on organic fertilizers. However, producing organic fertilizer was cumbersome and time consuming. Furthermore, organic fertilizers were hardly available in the input market.[5] The relatively higher yields of the Fair Trade certified farmers might be partly attributed to the Fair Trade cooperatives' extension training programmes and provision of equipment.

In Ethiopia, the certified farmers received 3.3 Ethiopian birr ($0.26) per kilogram for red cherry from their cooperative whereas the non-certified farmers received 3.21 birr ($0.25)[6]. The difference between average farm-gate price from cooperatives to the certified and the non-certified farmers is statistically insignificant, but the difference in prices paid by the private merchants to certified and non-certified farmers is significant. Nearly 75% of coffee harvest is sold to private traders from both certified and non-certified groups. While non-certified farmers receive an average of 6.41 birr ($0.50) per kilogram of sundried cherry from private traders, the same for certified farmers was 5.54 birr ($0.43).

In India, the local value chain of certified and non-certified farmers differs in structure and post-harvest management. The cooperative buys coffee in the form of red cherries from its members, while private traders buy coffee in dry processed form from any coffee farmer, certified or non-certified. Certified farmers sell most of their red coffee cherries (88%) to the cooperative and obtain an average farm-gate price of 18 Indian rupees (Rs) ($0.40)[7] per kilogram of red cherries. Considering that 6 kg of red cherries are needed to produce 1 kg of dry processed coffee, this is equivalent to around Rs109 ($2.42). The remaining 12% of the yield of the certified farmers is dry processed and sold to private traders for an average price of Rs94 ($2.1) per kg. The reason why certified farmers nevertheless sell parts of their coffee to private traders for prices lower than those paid by their own cooperative might be that they are in need for immediate cash at harvesting times, with rather low quality or processing demands. Unlike private merchants, the cooperatives only pay farmers after they have wet-processed their coffee cherries into coffee beans and received the returns from exporting them. On top of the price, the cooperative also provides dividends (premia) to their members at the end of profitable seasons. Thus, there are trade-offs for certified farmers in both these selling networks.

Farmers from all three types of certified channels in Nicaragua also sold their coffee both to their cooperative and to private merchants on the local markets. Organic and organic–Fair Trade cooperative members predominantly sold coffee to their cooperatives. This indicates that organic certified coffee farmers are more integrated in their cooperatives compared with the only Fair Trade certified farmers. Data also shows that organic and organic–Fair Trade farmers obtained significantly higher prices than the only Fair Trade and non-certified groups of farmers. The price differences between them and the Fair Trade farmers are not too great.

In the Ethiopian case study in Jimma Zone, the non-certified farmers achieved a higher net income from coffee than their Fair Trade certified counterparts. This goes against the general expectation that members of Fair Trade certified cooperatives earn more from their coffee than those in non-certified ones; however, the difference is not statistically significant.

For India, a comparison of net revenue earned from coffee per hectare of land shows that certified farmers earned 66% more than the non-certified farmers. While the income per unit of land for certified farmers is Rs10,956 (US$243), the non-certified farmers from the cooperative villages earned Rs7213 (US$160). This significant difference between the certified and non-certified groups can be explained by the fact that certified farmers received a significantly higher income from coffee on account of a strong price advantage in the certified value chain. In Nicaragua, the net revenues from coffee per unit of land reveal that organic and organic–Fair Trade farmers earned the highest gross margins i.e. 22,762 Nicaraguan córdoba (C$) (US$1012)[(8)] and C$22,337 (US$993) per year, respectively. The Fair Trade and conventional farmers earned C$19,682 (US$875) and C$19,079 (US$848), respectively. The higher net revenues earned by organic farmers compared with Fair Trade farmers demonstrates the high income-generating potential of organic certification.

14.4.3 The impact of certification on net income

This section provides empirical evidence on the income impacts of certification in the case studies in Ethiopia, India and Nicaragua based on the PSM method (Table 14.3). For Ethiopia, the mean differences are presented for both certified and non-certified groups after and before matching, respectively. Per capita income is statistically significant with a negative sign before the certified and non-certified farmers were matched, which infers that the latter group has a higher per capita mean income than the former group. However, it becomes statistically insignificant after matching. Further, in the case of total income, the result did not show any statistical difference either in the unmatched or in the matched samples.

For India, Table 14.4 shows the treatment effect differences between the treated group and the control group before and after the matching. It also shows the bias reduction that has been achieved after the matching. A perusal of the figures shows that the certified group has higher net income than the non-certified group. The difference in mean total income is in the order of Rs7392 (US$164.3) after matching. A reduction of 29% bias has been achieved by the matching and this result is robust to the extent where an unobservable factor may have caused the probability of drawing an observation from one group 1.6 times more than the probability of drawing an observation from the matched group. Similarly, the average per capita income is higher for the certified farmers by Rs6.23 (US$0.1) compared with the non-certified group. The reduction in bias is 14% and the robustness is in the order of 2.

In Nicaragua, the income impacts of certification are estimated using each of the certification

Table 14.3. Treatment effects from propensity score matching (PSM) in Ethiopia.

Variable	Sample size	Average treatment effect after matching	Average treatment effect before matching
Per capita income	246	–0.15 (1.10)	–1.36* (0.65)
Log total income	246	–0.18 (0.50)	–0.26 (0.26)

Standard errors are computed from clustered means. *Significant at 1% level.

Table 14.4. Treatment effect from propensity score matching (PSM) in India.

Outcome Variables	ATE[a] before matching	ATE[a] after matching	Bias reduction	Rosenbaum bound sensitivity analysis
Net income (Rs)	10,476***	7392***	29.4%	1.6
Per capita income (Rs)	7.26***	6.23***	14.2%	2

[a]Average treatment effect; ***significant at 1% level.

programmes as a treatment variable. Hence three PSM analyses are carried out between double certified and non-certified, Fair Trade certified and non-certified and finally, organic certified and non-certified groups.

Table 14.5 shows the PSM results for the double certified group using log total income and log net revenue from coffee as outcome variables. The three matching methods applied here (radius matching, kernel matching and 5-nearest-neighbour matching) show that double certified farmers have not earned any statistically significantly higher net total income compared with the non-certified farmers. However, for the second outcome variable, i.e. log of net revenue per hectare, radius matching and 5-nearest-neighbour matching methods reveal statistically significant coefficients. The average treatment effect is 0.571 for radius matching and 1.108 for 5-nearest-neighbour matching, meaning that double certified farmers have earned higher net revenue per hectare. The higher net revenue from coffee for the double certified farmers can be attributed to the higher farm-gate prices for them.

Table 14.6 gives the PSM results for Fair Trade certification and shows that there are no statistically significant coefficients for total income. As for the net revenue from coffee, only the radius matching has produced a statistically significant coefficient at 5% significance level. The other two matching methods have found statistically insignificant coefficients. Overall it can be inferred that there is no strong income impact from the Fair Trade network. Similarly, with organic certification there is no evidence of any income impacts as all the three matching methods have shown statistically insignificant coefficients for average treatment effect (Table 14.7).

Overall, the PSM results for different certification categories in Nicaragua show that there is no statistically significant impact on either double certified farmers or the organic and Fair Trade certified farmers when the total income is investigated. However, the double certified farmers have an edge over the non-certified farmers in terms of the net revenue per hectare. The former have earned higher net revenues, but these higher net revenues are not sufficient for them to secure a higher total income advantage. The reasons could be, firstly, that the margin of net revenue advantage is narrow and insufficient to make any difference in the total income and, secondly, the non-certified farmers may have other sources of income that compensate for their lower coffee income. Moreover, we have examined the net revenue from coffee per capita and it

Table 14.5. Propensity score matching (PSM) results for Fair Trade–organic double certification in Nicaragua.

Outcome variable	Matching methods	ATT	SE	T value
Log total income	Radius	0.032	0.191	0.17
	Kernel	–0.131	0.291	–0.45
	5-Nearest-neighbour	0.194	0.506	0.38
Log net revenue	Radius	0.571	0.325	2.68**
	Kernel	0.474	0.510	0.93
	5-Nearest-neighbour	1.108	0.413	2.68**

**Significant at 5% level.

Table 14.6. Propensity score matching (PSM) results for Fair Trade certification in Nicaragua.

Outcome variable	Matching methods	ATT	SE	T value
Log total income	Radius	–0.074	0.147	–0.50
	Kernel	–0.064	0.178	–0.36
	5-Nearest-neighbour	–0.071	0.176	–0.40
Log net revenue	Radius	0.494	0.216	2.29**
	Kernel	0.305	0.284	1.07
	5-Nearest-neighbour	0.313	0.273	1.15

**Significant at 5% level.

Table 14.7. PSM results for organic certification in Nicaragua.

Outcome variable	Matching methods	ATT	SE	T value
Log total income	Radius	0.136	0.221	0.61
	Kernel	–0.040	0.277	–0.15
	5-Nearest-neighbour	–0.006	0.026	–0.02
Log net revenue	Radius	0.315	0.247	1.27
	Kernel	0.367	0.404	0.91
	5-Nearest-neighbour	0.146	0.368	0.40

**Significant at 5% level.

was found that although double certified farmers received significantly higher farm-gate prices compared with the non-certified farmers, there is no significant difference in their net revenue per capita from coffee. The reason for this was found to be the household size. Double certified farmers on average have higher household sizes, making the net income per capita smaller.

14.5 Why Does Certification Have Differential Impacts?

The quantitative analyses in all three case studies show that coffee certification has either limited or no impact on the total household income of the participating coffee smallholders in Ethiopia and Nicaragua, while it has a positive impact in the Indian case study. It is worthwhile to discuss these findings qualitatively. In general, it is observed that all these certification efforts crucially depend on the performance of the primary cooperatives for their success. From our case studies, it is found that these cooperatives face severe challenges and constraints which are context specific and driven by individual historical, political, cultural and economic local conditions. However, some similarities can be found. Firstly, many primary cooperatives, especially in Ethiopia but also to a certain degree in Nicaragua and India, tend to follow rather top-down than participatory bottom-up approaches in their function. Being a member of a certified cooperative does not mean that a member is actively participating, nor that they understand the meaning and functionalities of certification. The formal educational level of many cooperative members is quite low, contributing to the fact that certification can be perceived as something 'abstract' and 'external'. Cooperative local officials tend primarily to communicate certification as a means to achieving higher income. However, achieving higher incomes through certification is a complex and long-term challenge. Coffee farmers are mostly in need of, and in favour of, immediate, direct and unconditional cash returns for their produce. While asking members of the certified cooperatives whether they knew that their cooperative was certified, in the Indian case study only 17% responded in the affirmative and the rest had no clue. Similarly, in the same case study, when the farmers were asked what they understood by certification, nobody could give a reasonable answer. In the Ethiopian case, only 33% of the members of certified cooperatives knew that their cooperative was certified. These examples can be taken as indications for lapses in local information, communication and participation. More research in this direction is needed.

Secondly, one of the key functions of farmer cooperatives is to provide a platform for horizontal relations where smallholders cooperate in the form of collective action in order to increase their revenues, and/or reduce costs and mitigate risks. In the above-mentioned three scenarios the members of cooperatives do not recognize themselves as active participants. The tendency is rather to consider the cooperative as a mere buyer of their product. For example, out of the 249 respondents in Ethiopia only 30% reported that they had a written contract. Some of them might have denied or forgotten that they actually did. It shows that, while the proportion of farmers with contracts in the certified cooperatives is relatively higher compared with non-certified cooperatives, their absolute number is small.

Thirdly, non-coffee-related services potentially provided by the primary cooperatives to their members play a different, and sometimes minor, role in all three cases. Credit delivery, for example, has the potential to be important since many coffee smallholders need to invest in various forms of upgrading in the production node, such as improving product quality, increasing volume and complying with certification standards (Elizabeth *et al.*, 2006). Our data show that only 23% of the respondents in Ethiopia said that they had obtained credit from their respective cooperatives. The figures are much higher in certified cooperatives: 32% of their members had received credit from their cooperative, while the figure was only 5% for members of non-certified cooperatives. In the Nicaraguan case, credit provision by the cooperative was far more frequent, with 87% of the interviewed respondents saying that they had received at least one credit from their cooperative. Also here members of certified cooperatives were more likely to receive credit from their cooperative (94% of the farmers in this group) compared with members of non-certified cooperatives (71% of the farmers in this group).

Although these shortcomings in the cooperative system are a common phenomenon in all the three case studies, there are some differences in the Indian case study. Despite the top-down approach in the Fair Trade certified cooperative in the Indian case, member farmers had been able to obtain relatively higher income mainly due to the fact that the case study region was historically an economically backward and tribal region where coffee was introduced as the only source of stable income. Hence, Fair Trade certification did provide a steady source of coffee income, which was non-existent earlier. Poor infrastructure and difficult geographical terrains in the Araku Valley had made it difficult for the tribal coffee farmers to transport their own produce to distant markets. The Fair Trade certified cooperative spared its members from carrying their sacks of coffee to the market; instead the cooperative bought coffee from the village centres. There were also some other benefits provided to the members, such as advance credit at the beginning of the planting season and subsidized harvest equipment.

The story of the cooperative movement is completely different in Nicaragua and Ethiopia. Coffee has been a way of life in both these countries and, as shown in Table 14.2, their allocation of cultivable land to coffee cultivation has been a major factor. Cooperatives in the coffee sector have existed for the past four decades in both countries. However, in spite of the long presence of cooperatives, the system still lacks dynamism.

14.6 Concluding Remarks and Recommendations

Livelihoods in developing countries face a big threat as the majority of the population in these countries depend on agriculture and practise traditional agricultural approaches. The popular notion is that certification of cash crops is a viable strategy as it helps to provide access to international markets and also helps in generating sustainable incomes. Based on our empirical case studies in Ethiopia, India and Nicaragua, the quantitative findings of this chapter show that certification on average has a differential impact in the three countries and this impact is highly case- and context-specific. Our findings in Ethiopia show that Fair Trade certified cooperatives have done far less to improve farmers' incomes and livelihoods. The major reason for this failure to increase incomes by Fair Trade certification is due to the lack of efforts by the cooperatives for capacity building. In the Indian case study, however, Fair Trade certification has positively affected farmers' income. Farmers who are members of Fair Trade certified cooperatives in India have received community-level benefits such as drinking-water installation, school uniforms for girls and some sports facilities. Since the level of the socio-economic parameters in the case study region such as physical infrastructure, assets of the households and education level had been quite low, the newly established Fair Trade certified cooperative could bring some of the early benefits through its Fair Trade marketing channels. More research needs to be done as to how some of these new cooperatives have been impacting farmers' livelihoods over the years in India.

In Nicaragua, farmers are traditionally organized in cooperative systems with rather bottom-up approaches. Fair Trade and organic coffee certification schemes as well as double certified schemes are common practice. Our findings in Nicaragua show that farmers in organic and organic–Fair Trade certified cooperatives receive comparatively higher farm-gate prices but also that these price advantages are mostly nullified by lower yields. Lower yields for organic farmers are a result of inadequate availability of organic production materials. Compared with Ethiopia, a significantly higher proportion of cooperative members in Nicaragua received credits from their cooperative.

In comparison, double certification in Nicaragua has given the cooperatives a distinct edge over single certification, since the net revenues have been higher in Fair Trade–organic cooperatives compared with single-certified cooperatives in Nicaragua. Overall, it is concluded that certification schemes, especially the double certified ones, can contribute to poverty reduction if complemented by efficient and frictionless local frameworks and by other effective development strategies. The first and probably most difficult challenge is the availability and maintenance of strong local organizational and infrastructural capacities. It is worth noting that the 'cooperative effect' needs to be considered while evaluating the impact of coffee certification on smallholders' livelihoods. Coffee cooperatives in these countries are challenged by a multitude of internal and external obstacles, making many of them weak actors. In order to implement certification programmes more efficiently and effectively, cooperatives as key local partners and vehicles for certification need to be substantially strengthened. The success of certification is hence often interlinked with capacities of the general agricultural sector in the respective country. Secondly, certification needs proper monitoring. The effective and regular verification of certification requirements by accredited inspectors is still challenging, not only in remote rural areas of developing countries.

Notes

(1) India also produces a large amount of Robusta coffee (CBI, 2010).

(2) There are, of course, differences in coffee quality and character.

(3) Fair Trade certification provides a social premium to the producer organizations that is invested in physical and social infrastructure and capacity building (www.fairtrade.net).

(4) Arunachal Pradesh, Assam, Manipur, Meghalaya, Nagaland, Mizoram and Tripura.

(5) Coffee production highly depends on fertilization. Many smallholders in Jinotea Municipality do not produce enough organic fertilization materials on their own farms. They do not have much livestock, for example, to produce enough manure for organic coffee farming. Markets for organic fertilizers are not developed.

(6) Using the exchange rate US$1 = 12.63 birr from 2009.

(7) Using the exchange rate US$1 = Rs45 from 2010.

(8) Using the exchange rate US$1 = C$22.5 from 2009.

References

Bacon, C. (2005) Confronting the coffee crisis: can fair trade, organic, and specialty coffees reduce small-scale farmer vulnerability in Northern Nicaragua? *World Development* 33(3), 497–511.

Beuchelt, T.D. and Zeller, M. (2011) Profits and poverty: certification's troubled link for Nicaragua's organic and fair-trade coffee producers. *Ecological Economics* 70(7), 1316–1324.

Castro, J.E. (2001) *Impact Assessment of Oxfam's Fair Trade Activities. The Case of COPAVIC*. Oxfam, Oxford, UK.

CBI (2010) *Coffee Statistics*. Coffee Board of India, Ministry of Commerce & Industry, Bangalore, Karnataka. Available at: http://www.indiacoffee.org/coffee-statistics (accessed 27 June 2017).

Chiputwa, B., Spielman, D.J. and Qaima, M. (2015) Food standards, certification, and poverty among coffee farmers in Uganda. *World Development* 66, 400–412.

Dempsey, J. and Campbell, R. (2006) *A Value Chain Approach to Coffee Production: Linking Ethiopian Coffee Producers to International Markets*. ACDI/VOCA Technical Bulletin. ACDI/VOCA, Addis Ababa, Ethiopia.

Elizabeth, D., Sebstad, J., Batzdorff, L. and Parsons, H. (2006) *Lessons Learned on MSE Upgrading in Value Chains: a Synthesis Paper*. AMAP BDS Knowledge and Practice microREPORT #71. US Agency for International Development (USAID), Washington, DC.

Fromm, I. and Dubón, J. (2006) *Upgrading and the Value Chain Analysis: The Case of Small-Scale Coffee Farmers in Honduras*. Paper presented at the Conference on International Agricultural Research for Development (Tropentag 2006), University of Bonn, October 11–13, 2006, Bonn, Germany.

Giovannucci, D. and Koekoek, F.J. (2003) *The State of Sustainable Coffee: a Study of Twelve Major Markets*. International Coffee Organization, International Institute for Sustainable Development and United Nations Conference on Trade and Development, Philadelphia, Pennsylvania.

Giovannucci, D. and Potts, J. (2008) *Seeking Sustainability: COSA Preliminary Analysis of Sustainability Initiatives in the Coffee Sector*. International Institute for Sustainable Development, Winnipeg.

Hopkins, R. (2000) *Impact Assessment Study of Oxfam Fair Trade*. Oxfam, Oxford, UK. Available at: http://www.fairtrade.net/793.0.html (accessed 5 October 2016).

ICO (2016) *Total production by all exporting countries*. International Coffee Organization, London. Available at: http://www.ico.org/prices/po-production.pdf (accessed 15 July 2016).

IICA (2004) *Cadena agroindustrial – Café*. Instituto Interamericano de Cooperación para la Agricultura, Ministry of Agriculture and Forestry (MAGFOR), and Japan International Cooperation Agency (JICA), Managra, Nicaragua.

Jena, P.R. and Grote, U. (2017) Fairtrade certification and livelihood impacts on small-scale coffee producers in a tribal community of India. *Applied Economic Perspectives and Policy* 39(1), 87–110.

Jena, P.R., Beyene Chichaibelu, B., Stellmacher, T. and Grote, U. (2012) The impact of coffee certification on small-scale producers' livelihoods: a case study from the Jimma zone, Ethiopia. *Agricultural Economics* 43 (2012), 427–438.

Jena, P.R., Stellmacher, T. and Grote, U. (2017) Can coffee certification schemes increase incomes of smallholder farmers? Evidence from Jinotega, Nicaragua. *Environment, Development and Sustainability* 19(1), 45–66.

Kamau, M.W., Mose, L., Fort, R. and Ruben, R. (2010) *The Impact of Certification on Smallholder Coffee Farmers in Kenya: The Case of 'UTZ' Certification Program*. Paper presented at the Joint 3rd African Association of Agricultural Economists (AAAE) and 48th Agricultural Economists Association of South Africa (AEASA) Conference, Cape Town, South Africa, September 19–23.

Muradian, R. and Pelupessy, W. (2005) Governing the coffee chain: the role of voluntary regulatory systems. *World Development* 33(12), 2029–2044.

Nelson, V. and Galvez, M. (2000) *Social Impact of Ethical and Conventional Cocoa Trading on Forest-Dependent People in Ecuador*. Natural Resources Institute, University of Greenwich, London.

OPM/IIED (2000) *Fair Trade: Overview, Impact, Challenges – Study to Inform DFID's Support to Fair Trade*. Oxford Policy Management / International Institute for Environment and Development, Oxford and London.

Pariente, W. (2000) The impact of Fair Trade on a coffee cooperative in Costa Rica. A producer's behaviour approach. Unpublished thesis, Université Paris I Panthéon Sorbonne, Paris.

Parrilli, M.D. (2001) *Subsector Analysis in the Mexican Craftwork: the Role of Fair Trade and Xochiquetzal in Tonolá and Chapala, Jalisco, and Olinalá. Guerrero*. Fair Trade Assistance, Culemborg, The Netherlands.

Ronchi, L. (2002) *The impact of fair trade on producers and their organizations: a case study with Coco-café in Costa Rica*. PRUS Working Paper No. 11. Poverty Research Unit at Sussex, University of Sussex, Brighton, UK.

Ruben, R. and Fort, R. (2011) The impact of Fair Trade certification for coffee farmers in Peru. *World Development* 40(3), 570–582.

Stellmacher, T. (2007) The historical development of local forest governance in Ethiopia. From imperial times to the military regime of the Derg. *Afrika Spectrum* 42(3), 519–530.

Tadesse, G., Zabaleta E. and Shennan, C. (2013) Coffee landscapes as refugia for native woody biodiversity as forest loss continues in southwest Ethiopia. *Biological Conservation* 169, 384–391.

Tiyapongpattana, T.-A. (2001) *Fair Trade Effects Studies 2001 – Thailand*. Fair Trade Assistance, Culemborg, The Netherlands.

UCA Soppexcca (2011) *Jinotega, the Capital of Coffee in Nicaragua*. Available at: http://www.soppexcca.org/en/jinotega.html (accessed 5 November 2016).

Utting-Chamorro, K. (2005) Does Fair Trade make a difference? The case of small coffee producers in Nicaragua. *Development in Practice* 15(3/4), 584–599.

Valkila, J. (2009) Fair Trade organic coffee production in Nicaragua – sustainable development or a poverty trap? *Ecological Economics* 68, 3018–3025.

Valkila, J. and Nygren, A. (2009) Impacts of Fair Trade certification on coffee farmers, cooperatives, and laborers in Nicaragua. *Agriculture and Human Values* 27(3), 321–323.

Wooldridge, J.M. (2002) *Econometric Analysis of Cross-section and Panel Data*. MIT Press, London.

15 The Relevance of Reliability, Reputation and Respect for Producer-level Benefits of Organic and Fair Trade Certification for Smallholders

Linda Kleemann*
Research Area Poverty Reduction, Equity and Development, Kiel Institute for the World Economy

15.1 Introduction

Access to international agricultural markets is often linked to certification requirements. For horticultural products, certification with GlobalGAP is a market entry condition for conventional food and organic or Fair Trade certification is required for the high-value organic food markets. Retailers normally require that their suppliers adhere to one or more such standards (Henson *et al.*, 2011). The literature shows that this can be seen both as an entry barrier for smallholders (Schuster and Maertens, 2013) and as an agent for change and innovation (Warning and Key, 2002; Maertens and Swinnen, 2009; Miyata *et al.*, 2009; ITC, 2011; Bellemare, 2012; Subervie and Vagneron, 2013). However, in many cases certifications fail after the first or second year and hence short-run benefits blow off quickly (Bellemare, 2012). Considering the large initial investment required, the existence of long-term net positive effects crucially depends on survival rates, i.e. the length of a specific contract farming relationship or certification period. This research tries to shed light on the reasons for the failure of some and long-run success of others.

Due to the high certification costs, smallholders are usually not able to pay for these costs themselves. In a typical setting where smallholders are involved, the buyer and/or a third party, usually a donor or a non-governmental organization (NGO), co-finances the certification costs. When the buyer co-finances the certification, this is usually linked to contract farming. The literature finds positive short-run income effects for this kind of contract farming (Bolwig *et al.*, 2009; Barrett *et al.*, 2012; Bellemare, 2012), when successful. However, there are huge differences between countries, value chains and firms (see e.g. Swinnen and Kuijpers, 2016 for a summary of the evidence).

While differences between countries and value chains may (in part) be explained through differences in market structure, regulation and market dynamics, differences within the same country and value chain need to be explained by internal factors.

In this chapter we develop a conceptual model of farmer–business relationships under certification and highlight the main arguments using empirical data. In particular, we show that signalling reliability and building up a reputation

*E-mail: Linda.Kleemann@ifw-kiel.de

of trust and respect increase the likelihood of successful contracts. Based on data from 386 either GlobalGAP or organic certified smallholders in the pineapple export sector in Ghana and from 257 organic or Fair Trade certified smallholders in Uganda, we show that there are large differences between firms with respect to the success of their contract farming relationships. In addition, we use subjective self-statements to gain deeper information about the farmers' perceptions and motivations, as these shape expectations. The match or mismatch between farmer motivation to join a certification or contract arrangement and the perception of the outcome of this process defines whether the farmer will be satisfied or disappointed with the outcome. Non-alignment may explain many failures of contract schemes. Secondly, alignment between firm management practices and farmer behaviour shapes satisfaction and long-run success. Our case studies show that three Rs – reliability, reputation and respect – constitute the basis for contract relationships that benefit all. Standards linked with contracts are short-run agents of change; individual farmers and firms determine whether they translate into long-run benefits.

The rest of the chapter is organized as follows: Section 15.2 describes the conceptual model; Section 15.3 presents the empirical data; and Section 15.4 concludes.

15.2 Conceptual Model

Agricultural contracts in developing countries are often faced with weak contract enforcement translating into high transaction costs. Certifications additionally raise contracting costs due to the initial investment they require. When both sides invest in the farmer–buyer relationship, cheating becomes costly. Under these conditions, even beneficial contracts for both sides may not take place, because benefits manifest only in the long run. Signalling commitment and building up a good reputation on both sides can lower enforcement costs and thus increase the likelihood of a contract. In addition, donors or NGOs, by taking over certification costs, can lower contracting costs. Figure 15.1 shows the stylized set-up for a farmer–buyer relationship with certification.

Agrobusinesses are attractive for smallholders because they potentially offer high prices and a relatively stable market, whereas smallholders may offer a secure supply base to agrobusinesses. Both contribute to certification with effort (smallholders) and knowledge and (part of the) cost (agrobusinesses), respectively. The contract setting can be described as an interlinked contract (Swinnen and Kuijpers, 2016).(1) With weak external contract enforcement, contracts need to be self-enforcing. Certification may aid self-enforcement when it forces both parties to invest prior to gaining from the contract, since each party wants to recover the sunk cost of the investment.

From the farmers' perspective, side-selling to other buyers is the most frequently reported way of breaching a contract (Swinnen and Kuijpers, 2016). Buyers, on the other hand, may breach contracts by failing to uphold promised volumes or prices or by cheating in weighting or quality assessment of the product (Cungu *et al.*, 2008; Barrett *et al.*, 2012; Mujawamariya *et al.*, 2013; Saenger *et al.*, 2014; Gelaw *et al.*, 2016).

Gelaw *et al.* (2016) showed how in such a setting the reliability and credibility of the buyer, in this case a trader, matters for marketing preferences of farmers. Farmers aim to establish long-term relationships with reliable buyers in order to lower transaction costs. Tadesse and Shively (2013) found that farmers tested their buyers for trustworthiness.

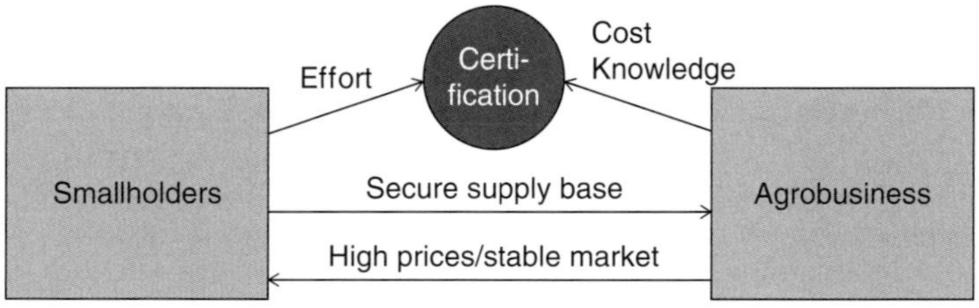

Fig. 15.1. Stylized farmer–buyer relationship with certification.

To explain this formally, certifications may be a situation in which individually over the short run, the net surplus (Π) of certification for both farmers (*f*) and buyers (*b*) is negative, i.e.:

$$\Pi_{f1} = p_{cf1}q_{cf1} - p_{wf0}q_{wf0} - I_f - \Delta c_{cf}q_{cf} < 0 \quad (15.1)$$

and

$$\Pi_{b1} = p_{cb1}q_{cb1} - p_{wb0}q_{wb0} - I_b - \Delta c_{cb} < 0 \quad (15.2)$$

for one season (in this case season 1) with p_{ci} and p_{wi} being the prices achieved without (*w*) and with (*c*) certification, q_{ci} and q_{wi} being the respective volumes, I_i the investment in certification and Δc_{cb} the cost difference with and without certification[(2)] for $i = f, b$.

However, over a certain number of seasons it becomes positive:

$$\sum_{t=1}^{j} \Pi_{fi} = \sum_{t=0}^{j} p_{cij}q_{cij} - p_{wij}q_{wij} - I_i - \Delta c_{ci}q_{ci} > 0 \quad (15.3)$$

for $i = f$ and $i = b$ individually. The time span required depends on I_i, Δc_{ci}, p_{cij} and q_{cij} and may be different for farmers and buyers. After that the net benefit rises with each added season. Under perfect information over the costs and benefits at the time of certification and in the future and without added external risk (market volatility, weather and climate, etc.), this contract would be self-enforcing. However, in the absence of perfect information and under weak contract enforcement, this turns into a prisoner's dilemma situation in which signalling reliability increases the chances of contract success. In the next section we analyse such signalling practices from the side of the firms.

15.3 Empirical Results

Two separate datasets are used to exemplarily highlight the factors analysed in this chapter. The first dataset comes from a survey of pineapple farmers in Ghana in 2010 and the second from a survey of Ugandan pineapple farmers in 2008. Both datasets include farmers who sell to different export-oriented agrobusinesses, but also sell part of their harvest on the local market without any reference to the certification.

The farmer survey in Ghana was conducted in the Central, Eastern and Greater Accra regions in a radius of about 100 km north and west of Accra, which constitutes the main pineapple-growing area. In Uganda, the survey was conducted in the southern, eastern and central regions, also the main pineapple-growing area. In both cases random samples were drawn from the lists of GlobalGAP, organic or Fair Trade certified smallholders. All of them sell to agrobusinesses that in turn export their products fresh or processed mainly to Europe. The questionnaire contained questions on production, certification and marketing, but also household characteristics and social capital. The dataset from Uganda includes 257 farmers with organic (121 farmers) or Fair Trade (163 farmers) certification for their pineapple farms. The dataset from Ghana includes 386 farmers from 75 villages with either GlobalGAP (201 farmers) or organic (185 farmers) certification for their pineapple farms. Table 15.1 presents an overview of the data.

Both datasets are similar in terms of age, gender and household size. The vast majority of pineapple farmers are male, on average around 45 years old and with about 12 years of pineapple farming experience. Higher primary completion levels in Ghana coincide with higher literacy levels among the surveyed farmers. Pineapple farms in Ghana are slightly larger than in Uganda, with 5.71 versus 4.67 acres (2.3 versus 2 ha). These figures do not include fallow land and land used for crops other than pineapple. Thus total farm sizes are larger. The pineapple varieties planted are MD2, Smooth Cayenne, Queen and Sugarloaf (the latter only in Ghana), with some farmers planting two varieties.[(3)]

In both countries, the dataset covers different exporters involved in contract farming. In Uganda, the survey covers three different exporters for fresh and dried pineapple. These include Amfri Farms, Bio Uganda and Biofresh.

In Ghana, a larger variety of exporters are covered. In addition, data are available from a detailed case study of the agrobusiness Blue Skies, a large-scale export-oriented processor in Ghana. The data sources are linked through farmers identified in the survey and in the case study. In the Ghana sample, 71 farmers can be linked to Blue Skies. We are hence able to compare farmers producing for Blue Skies with farmers in the same sector but producing for another

Table 15.1. Descriptive statistics of selected variables

Variable	Definition	Ghana mean (*N* = 386)	Uganda mean (*N* = 257)
GENDER	Gender of farmer (1 if male, 0 otherwise)	0.94	0.81
AGE	Age of farmer	44.71	45.74
HHSIZE	Household size (persons living in household)	5.71	4.67
PRIMEDU	Share of farmers that completed primary education	0.88	0.59
FSIZE	Pineapple farm size (acres)	5.53	3.12
OWNLAND	Share of land owned	0.29	0.76
CREDIT	Access to credit during the last 5 years (1 if yes, 0 otherwise)	0.28	0.58
WEALTH	Number of durable goods owned	6.76	4.92
EXPER	Years of experience in pineapple farming	11.49	12.86
DIST	Distance to the closest local market (hours)	0.75	0.45
ASSIST	Assistance or training for farming received during last year (1 if yes, 0 otherwise)	0.82	0.59
ORGANIC[a]	Organically certified (1 if yes, 0 otherwise)	0.52	0.78
GGAP	GlobalGAP certified (1 if yes, 0 otherwise)	0.48	0.12
FT	Fair Trade certified (1 if yes, 0 otherwise)	0.13	0.44
CERTYEAR	Number of certified years	3.46	2.65
BS	1 if buyer is Blue Skies, 0 otherwise	0.20	

[a]In Uganda, 21 farmers have both organic and Fair Trade certification. In Ghana the overlap between organic and Fair Trade is 11. There is no triple certification (organic, Fair Trade and GlobalGAP).

firm. The case study of Blue Skies was based on interviews with its suppliers, i.e. farmers, employees, management and communities in which Blue Skies was active in 2013.[(4)] For the purpose of this chapter, only the farmer and management interviews will be used. Set up in Ghana in 1998, Blue Skies produces fresh cut fruit and fruit salads for export mainly to Europe and freshly squeezed juice for the local market. It buys both organic and GlobalGAP certified fruit from local farmers, mostly in a contract farming arrangement, but occasionally also on the spot market. Contracts with farmers are renewable annually and specify the certifications, crop variety, quality, ripeness and terms of payment. Prices are fixed in £ sterling and renegotiated annually. The overall acreage grown for Blue Skies is 1928 (including own farms).

For the detailed case study, a combination of a standardized questionnaire and open qualitative interviews using the most significant change technique were used. Ten farmers were interviewed, representing the major crops grown for Blue Skies by smallholders: pineapple (four farmers), mango (three), papaya (three) and coconut (one). They were randomly drawn from the contracted suppliers list. One person from the Blue Skies agronomy team always introduced us. Farmer interviews were therefore not conducted confidentially. This was the only possible way to be well received. Farmers were nevertheless very opinionated and sometimes even asked the agronomy team to listen and witness their complaints or requests. All interviews were made on the farms and with the farmers themselves, all of whom were male. Each interview took 30–60 min. In addition, interviews and informal discussions were led with the management of Blue Skies throughout the study period. The management of Blue Skies provided us with all the information requested on management practices, farmer statistics and policies, extension, certification, etc. and was available for clarifications and feedback.

15.4 Analysis

15.4.1 Signalling commitment: certification as an agent of change in farmer–agrobusiness relationships

Farmers in Ghana and Uganda were asked about the existence and relevance of contract farming.

While only 28% of farmers in Ghana were producing under contract before certification, it was 100% after certification. In Uganda, the figures are 14% before and 89% after certification. For these examples, certification and contract farming go hand in hand and contracts signal longer-term commitment from both sides.

In both cases, farmers and firms reported that the certification required high initial investments from both sides (farmer and expert/buyer). For the farmers, this was mainly effort for training and changes required in production. In many cases production costs rose as a consequence of the certification requirements. For the buyers, this was the time invested and financial resources spent on the certification process. In some cases the buyer paid for the whole cost of the initial certification, in some cases NGOs, donors and/or farmer groups themselves co-financed the certification process. The consequence of these investments was an increase in the salience of the farmer–buyer relationship, which manifested itself – amongst others – in contracts. These contracts were for 1 year in over 90% of the cases and stated a fixed price or a minimum price to be paid, as well as the obligation to give the particular buyer priority over all others. In some cases a guarantee on the volume of sales at a given quality was also given. Both sides had the desire to force the other party to stick to the relationship at least until the costs of certification were recovered. In addition, in some cases the buyer held the certificate (i.e. the certificate had the buyer's name on it). This result of certification can be best understood in light of the fear of side-selling by farmers, as summarized in Swinnnen and Kuijpers (2016). Focus group discussions in Uganda and the case study in Ghana revealed that many farmers also feared non-buying by agrobusinesses. In fact, in the detailed case study farmers mentioned reliability and consistency and volume as the most important positive points of their contract farming relationship, giving these points considerably more importance than prices.

In addition to these formal changes, self-reported changes include a more intense relationship (farmer and agrobusiness talking more often to each other) and an improved overall relationship following certification. Specifically, 67% of farmers reported an improved overall relationship, against 31% reporting no change and 2% a worse relationship; and the results for the intensity of the relationship were 79%, 20% and 1%, respectively.

In addition, 77% of the farmers reported that they sold for more years to the same buyer compared with their peers and compared with the situation before certification. The certification process can take several years (e.g. 3 years for organic certification, including the transition period from chemical to organic inputs), hence a considerable time lag between the decision and the first market transaction as a certified vendor. Longer relationships (selling to the same buyer for more years) allow for renewing the certification and recovering the initial investment. This is in the interest of both sides, due to the high initial investment.

Thus, certification not only alters prices and costs; it is also a driver of change in farmer–agrobusiness relationships and contract specification.

15.4.2 Signalling the right kind of benefits: expectation management towards certification

We find that prior expectation towards the potential benefits of certification likely influences farmers' motivations to become certified and thus their level of satisfaction with the certification afterwards. Firstly, we asked all farmers directly what their motivations for organic, Fair Trade and GlobalGAP certification were.[5]

Those informed by NGOs or donors[6] stated far more often than those informed by agrobusinesses that environmental concerns, health or food safety were determining factors, whereas those informed by agrobusinesses considered market access, buyer demands and contracts were the most important (Table 15.2). But there are differences between certifications: organic exhibits a more similar pattern between different information sources compared with Fair Trade and GlobalGAP. Generally, buyers and other farmers stress 'hard' information (on prices, markets and yield) on certification, whereas NGOs and donors put the focus on 'soft' information such as environmental hazards and safe handling. This is the outcome from feedback discussions with buyers and donors and is also reflected in the certification training material provided

Table 15.2. Stated motivation to become certified (pooled across both countries).

	GlobalGAP		Organic		Fair Trade	
	Informed by business	Informed by NGO or donor	Informed by business	Informed by NGO or donor	Informed by business	Informed by NGO or donor
Better yields	10%	18%	9%	17%	4%	11%
Better prices	16%	9%	26%	31%	17%	56%
Health or food safety reasons	26%	10%	11%	7%	2%	2%
Environmental concerns	9%	37%	16%	26%	13%	13%
Better contracts with exporters	15%	4%	13%	3%	30%	4%
Customer demands	16%	0%	6%	3%	34%	1%
Market access	7%	19%	18%	6%	1%	13%
Cultural reasons/ tradition	0%	1%	0%	5%	1%	1%
Other reasons	1%	2%	1%	2%	0%	0%

by these groups. Even more, it is also mirrored in the personal motivation that farmers stated for aiming at a particular certification.

Next, we asked farmers whether their expectations were satisfied (Table 15.3). There is a strong mismatch between satisfaction levels and expectation for both GlobalGAP and Fair Trade certification. Businesses seem to be better at managing expectations. According to our data, the quality of the relationship between farmer and agrobusiness, as subjectively perceived by the farmer, is significantly better in those cases where the agrobusiness provided the initial information (test statistic of two-sided *t*-test is 2.77). In addition, when the agrobusiness provided the initial information, it likely also organized the certification process and even paid for it (correlation of 0.67). Also, in those cases where the exporter paid for the certification, farmers received on average double the amount of training than in all other cases. The quality of the relationship was also perceived as significantly better when the agrobusiness organized the certification process (test statistic of two-sided *t*-test is 6.11).

This result points towards the importance of integrating and engaging both partners early in the certification process in order to align expectations and build trust. This was confirmed in the discussions with both sides, with the key factor being building mutual trust. Disappointments due to wrong expectations or unreliability of the other party were mentioned as the main reasons for failures of contract schemes. This means that the ability of both certification and contract to deliver on the expectations they created will determine their success.

There are large differences in how well this works between firms, as the detailed case study below shows.

15.4.3 Signalling reliability: analysis of the firm factor

Long-term contract relationships allow for positive partner-specific investments on both sides such as on farm pack houses, planting of specific varieties, etc. Contracts last when both sides stick to each other. Even though sticking with the same farmers would be beneficial for firms as shown in the conceptual model, firms differ widely in their success in actually doing so, as shown in Table 15.4. We argue that signalling reliability and building up a reputation of trustworthiness by sticking to agreements, transparency and sharing of benefits and risks are key factors in successful certification and contract management with small-scale farmers under weak contract enforcement settings.

Blue Skies in Ghana is a firm that manages these relationships particularly successfully. Table 15.5 shows that Blue Skies farmers are

Table 15.3. Satisfaction with expectations towards certification (pooled across both countries).

	GlobalGAP		Organic		Fair Trade	
	Informed by business	Informed by NGO or donor	Informed by business	Informed by NGO or donor	Informed by business	Informed by NGO or donor
Yes	33%	10%	29%	24%	77%	23%
Partly	56%	12%	59%	48%	22%	64%
No	11%	78%	12%	28%	1%	13%

Table 15.4. Differences between firms in the same sector in Uganda.

Variable	Firm A	Firms B and C	*t*-Stat.
Average length of contract relationship (years)	3.2	1.8	1.71*
Frequency of contract breaches in last 3 years (number)	2.1	6.8	6.71***
Farmer satisfaction with buyer (1 = totally satisfied, ..., 5 = not at all satisfied)	2.0	3.4	3.01***

Significance levels: *10%; **5%; ***1%. Firm names are not revealed, in order to guard their confidentiality.

Table 15.5. Blue Skies farmers in comparison with other contract farmers.

Variable	Definition	Blue Skies farmers (*N* = 71)	All other farmers (*N* = 282)	*t*-Stat
ROI	Return on investment in pineapple farming (1 year)	3.13	1.90	3.56***
PRODCOS_KG	Production cost/kg fruit	0.11	0.12	0.63
REV_KG	Revenue GHG/kg fruit sold	0.26	0.17	9.17***
PROFIT_KG	Profit GHG/kg fruit sold	0.15	0.06	5.21***
YEARS_BUYER	Number of years already selling to same buyer	2.42	0.97	10.62***
REL_BUYER	Quality of relationship to buyer on scale from 1 (very good) to 4 (very bad)	1.31	2.34	10.66***
Details of the quality of the relationship between buyer and seller:				
PICKUP_BUYER	1 if satisfied with delivery/pickup arrangements, 0 otherwise	0.94	0.36	8.24***
VOL_BUYER	1 if satisfied with volumes bought by buyer, 0 otherwise	0.76	0.15	7.00***
BUY_GUARANT	1 if guaranteed volumes bought, 0 otherwise	0.83	0.26	10.08***
TIME_PAY	Time lag from pickup to payment (1 = same day, ..., 5 = 3 months or more)	2.91	3.14	2.25**
Intensity of the relationship between buyer and seller:				
MEET_BUYER	Frequency of meetings between buyer and seller (times/year)	10.65	5.96	5.38***
PHONE_BUYER	1 if phone number of buyer known, 0 otherwise	0.68	0.20	5.09***

Significance levels: *10%; **5%; ***1%. We use a conversion factor of 1 Ghana cedi (¢) = €0.46 (calculated on the basis of the exchange rate on 12 January 2010).

better off in several respects. They have a higher return on investment and a higher profit that is determined by higher revenues, not by lower costs.

Blue Skies farmers are also better off in another respect, which was identified as a crucial success factor in the previous section. Farmers producing for Blue Skies state that they have a significantly better and more intensive relationship with their buyer, than all other farmers in the sample. This relationship also lasts longer, on average more than 2 years instead of less than 1 year.(7) Intensity is measured by whether or not they have the phone number of the buyer, the frequency of meetings, and how much they know about the further use of their pineapple (destination, processing). Whether the better relationship results in better economic outcomes or vice versa remains unclear. These correlations might not be causal. Therefore we have tried to verify these results through qualitative interviews.

In addition, selection for the 'better' farmers may be an issue that affects some of these results. We tested some standard measures such as farm size, production costs and experience and while there are no significant differences between Blue Skies farmers in these respects, there may be other factors that we did not capture such as fruit quality and farmer reliability. The selection process is one of the main aspects we focused on in the qualitative interviews.

All agrobusiness firms cluster in a small area. They target the same regions for small-scale farmers supplying them with fresh produce. The case study revealed that, because Blue Skies has a good reputation as buyer, they might have the first choice in terms of farmers wishing to supply to them. We asked the farmers directly in order to find out whether happier and, in economic terms, better farmers selected into contract farming with Blue Skies or whether farming for Blue Skies made them better off and happier afterwards. As most farmers had experiences with other agrobusiness firms, they were able to compare. Farmers considered the secure and reliable long-term market and payment stream that Blue Skies is providing as the most important impact channel, especially when comparing with other buyers. Blue Skies is respected for its corporate culture of respect, social equality and openness up to the point that farmers imitate it themselves (Table 15.6). Several farmers mentioned their admiration for Blue Skies' management, especially related to mastering past market challenges. Blue Skies is also respected for the quality of its advice and training to the farmers, which is, compared with others, much more targeted to their needs and also for taking up farmers' suggestions and ideas (Table 15.6). In addition, Blue Skies, in partnership with two of its buyers, supports community projects through a foundation. Projects are proposed by the farmers and owned by the communities. The interviews showed that the foundation is an important add-on because it gives Blue Skies and its farmers a good standing in the communities, again building up a good reputation. Although these projects are judged as important by the farmers, they are nevertheless second to a stable market (Table 15.6).

The positive management of problems and difficulties by Blue Skies was particularly mentioned by several farmers. They are supporting farmers in testing new farming techniques, e.g. using plastic mulch, while being careful not to induce high expectations that they cannot meet. While the farmer is not happy with the low demand, he understands the demand situation and respects Blue Skies for its efforts.

We conclude that the satisfaction of Blue Skies farmers and their economic success is due, at least to a considerable part, to the way Blue Skies treats its farmers and not due to selection effects. But does this also benefit Blue Skies, i.e. is it a win–win situation? Without being able to establish causality, we observe that Blue Skies has had its operations in Ghana since 1998, over time considerably increasing in size. During this period, many others have failed or remained much smaller (e.g. Peelco, WAD, Coastal Groves, Kingdom Fruit Juice, Nsawam Cannery, Athena).

As a next step, we try to understand the corporate causes behind the big difference between Blue Skies farmers and other contracted farmers that we found in the farmer survey and in the qualitative interviews. While Blue Skies does not differ significantly from other agrobusinesses in terms of buying practices and corporate social responsibility, it differs in terms of soft factors. Reliability and consistency, corporate culture and training were identified as most important impact channels by the farmers (Table 15.6). We thus contoured the main factors within Blue Skies by reviewing their policies and observing their actual behaviour in day-to-day business. This created a strong identification with Blue Skies among farmers and hence led to lower contract breaches. In addition, Blue Skies behaves in a protective way towards its farmers

Table 15.6. Most important channels from the perspective of farmers, in descending order (interviews conducted by the author).

Subject	Response
Reliability and consistency	'Blue Skies is the most reliable buyer and always pay everything and on time. Prices are fair and we are told about quantities in advance. And there are additional incentives that other buyers do not provide.'
	'Sometimes we expect to sell more but we understand it is because of the orders that Blue Skies receives from their customers. We stay with Blue Skies because it is reliable and we can constantly supply them. There is no other consistent buyer in Ghana.'
	'I would prefer to sell to Blue Skies even if I get a higher price elsewhere.'
	'Other exporters were not reliable.'
Volume	'We sell almost all our fruit to Blue Skies. And we would sell more. We want to expand the farm and improve housing for workers if we are able to sell more.'
Corporate culture	'We feel that we are all part of the Blue Skies family. We can openly discuss our problems and complaints with the agronomy team. Aspiring farmers are built up to succeed by Blue Skies. We admire how they manage, especially in difficult times.'
Training	'The constant training from Blue Skies is very beneficial. They visit us every 2–6 weeks for audits and trainings that cover amongst others certification, cropping, farm management. We also ask for advice with current prevalent problems. They take our concerns seriously.'
Credit	'We would like to receive a loan for the expansion of the farm. We cannot get it from Blue Skies and the banks are not helping either. They have very high interest rates and demand huge collateral. But on an individual basis, needs are considered. We can get soft loans (without interest) as advance payment. We know that we can count on Blue Skies that they will do their best.'
Community projects	'I was very involved in getting the Foundation project in my community. I am now also in the management committee.'
	'We are applying to the Foundation to get a Junior High School (JHS) to our community. But more important is more demand for fruit.

and surrounding communities, trying to buffer them against market volatilities, while transparently communicating its own challenges. This combination of protection and open communication creates a trustworthy and resilient relationship between farmers and Blue Skies. In summary, our case study shows that Blue Skies has managed to build up, signal and maintain a reputation of reliability and respect.

15.5 Conclusion

With increasing relevance of certification standards such as GlobalGAP, organic and Fair Trade and associated contract relationships between exporters and smallholders, many researchers have analysed the income and welfare effects of such arrangements. But they have so far neglected the role of certification as a structural element driving contract outcomes. This chapter shows that certification is an agent of change in farmer–agrobusiness relationships by developing a conceptual model in which certification enters as a cost factor. This cost factor creates a considerable time lag between the decision to invest and the first benefits, which in turn provokes a situation in which contracts are only beneficial if they are sustained over a minimum period of time. In order for such contracts to exist, signalling reliability, aligning expectations of farmers and buyers (i.e. agrobusinesses) and building trust between the partners is crucial for the success of the whole process. Some agrobusinesses are more successful than others in managing the required kind of trustful and strong relationship with their contracted smallholders. This means that behaviour of the individual firm matters more than has been taken into account in previous research. A detailed case study of one successful firm concludes that three 'Rs' – reliability, reputation and respect – constitute the basis for contract relationships that benefit all.

Future research would benefit from calculating survival rates of agrobusiness–smallholder contracts and linking them to economic benefits. For supporters of certification processes, be it NGOs, donors, or agrobusiness firms, this means that more importance should be placed on longevity of contracts and on 'soft' factors such as trust building and forming a joint culture in addition to 'hard' facts such as market opportunities and requirements.

Notes

[1] Bell and Srinivasan (1989) defined interlinked market transactions as a transaction in which the parties trade in at least two markets on the condition that the terms of all trade between them are jointly determined.

[2] Including re-certification cost.

[3] The farmer survey from Ghana has been quantitatively analysed in particular with respect to return of investment in certification and agricultural practices in Kleemann *et al.* (2014) and Kleemann and Abdulai (2013). Here, we focus on the subjective statements that farmers were asked to give during the survey.

[4] The data gathered for the case study were collected by the author for a report commissioned by Waitrose, one of the buyers of Blue Skies products. The information is used with permission from Blue Skies and Waitrose.

[5] By stressing differences between statements and actions, they affirm that believing in what people say can be misleading (Manski, 2004). Consequently, one rarely sees subjective data in empirical papers. We want to break with this tradition and compare our quantitative results with qualitative information about farmers' own statements on the subject. We are aware of the measurement errors that come with individual differences in interpretation of questions and expressions, and thus cannot make accurate statements using this methodology.

[6] Government extension services are hardly ever (in 9% of all cases) relevant information providers.

[7] The sample was random and representative at the time of survey, which implies that there should be no differences in average contract duration if there is no 'firm factor'.

References

Barrett, C., Bachke, M., Bellemare, M., Michelson, H., Narayanan, S. and Walker, T. (2012) Smallholder participation in contract farming: comparative evidence from five countries. *World Development* 40(4), 715–730.

Bell, C. and Srinivasan, T.N. (1989) Interlinked transactions in rural markets: an empirical study of Andhra Pradesh, Bihar and Punjab. *Oxford Bulletin of Economics and Statistics* 51(1), 73–83.

Bellemare, M.F. (2012) As you sow, so shall you reap: the welfare impacts of contract farming. *World Development* 40(7), 1418–1434.

Bolwig, S., Gibbon, P. and Jones, S. (2009) The economics of smallholder organic contract farming in tropical Africa. *World Development* 37(6), 1094–1104.

Cungu, A., Gow, H., Swinnen, J.F.M. and Vranken, L. (2008) Investment with weak contract enforcement: evidence from Hungary during transition. *European Review of Agricultural Economics* 35(1), 75–91.

Gelaw, F., Speelman, S. and Van Huylenbroeck, G. (2016) Farmers' marketing preferences in local coffee markets: evidence from a choice experiment in Ethiopia. *Food Policy* 61, 92–102.

Henson, S., Masakure, O. and Cranfield, J. (2011) Do fresh produce exporters in sub-Saharan Africa benefit from GlobalGAP certification? *World Development* 39(3), 375–386.

ITC (2011) *The Impacts of Private Standards on Producers in Developing Countries*. International Trade Center Literature Review Series on the Impacts of Private Standards, Part II. ITC, Geneva.

Kleemann, L. and Abdulai, A. (2013) Organic certification, agro-ecological practices and return on investment: evidence from pineapple producers in Ghana. *Ecological Economics* 93, 330–341.

Kleemann, L., Abdulai, A. and Buss, M. (2014) Is organic farming worth its investment? The adoption and impact of certified pineapple farming in Ghana. *World Development* 64, 79–92.

Maertens, M. and Swinnen, J.F.M. (2009) Trade, standards, and poverty: evidence from Senegal. *World Development* 37(1), 161–178.
Manski, C.F. (2004) Measuring expectations. *Econometrica* 72(5), 1329–1376.
Miyata, S., Nicholas, M. and Dinghuan, H. (2009) Impact of contract farming on income: linking small farmers, packers, and supermarkets in China. *World Development* 37(11), 1781–1790.
Mujawamariya, G., D'Haese, M. and Speelman, S. (2013) Exploring double side-selling in cooperatives, case study of four coffee cooperatives in Rwanda. *Food Policy* 39, 72–83.
Saenger, C., Torero, M. and Qaim, M. (2014) Impact of third-party contract enforcement in agricultural markets – a field experiment in Vietnam. *American Journal of Agricultural Economics* 96(4), 1220–1238.
Schuster, M. and Maertens, M. (2013) Do private standards create exclusive supply chains? New evidence from the Peruvian asparagus export sector. *Food Policy* 43, 291–305.
Subervie, J. and Vagneron, I. (2013) A drop of water in the Indian Ocean? The impact of GlobalGAP certification on lychee farmers in Madagascar. *World Development* 50, 57–73.
Swinnen, J. and Kuijpers, R. (2016) *Value Chain Innovations for Technology Transfer in Developing and Emerging Economies: Concept, Typology and Policy Implications*. Discussion Paper 376/2016, LICOS Discussion Paper Series. LICOS Centre for Institutions and Economic Performance, Leuven, Belgium. Available at: http://ssrn.com/abstract=2771612 (accessed 12 February 2016).
Tadesse, G. and Shively, G. (2013) Repeated transactions in rural grain markets of Ethiopia. *Journal of Development Studies* 49(9), 1172–1187.
Warning, M. and Key, N. (2002) The social performance and distributional consequences of contract farming: an equilibrium analysis of the Arachide de Bouche program in Senegal. *World Development* 30(2), 255–263.

16 The Way Forward

Priyanka Parvathi[1]*, Ulrike Grote[2] and Hermann Waibel[1]
[1]Institute of Development and Agricultural Economics, Leibniz University Hannover, Germany; [2]Institute for Environmental Economics and World Trade, Leibniz University Hannover, Germany

Fair Trade and organic markets have both steadily grown in recent decades but are they considered and perceived as a winning combination? This book has presented case studies on consumers and producers of both Fair Trade and organic products. Some major outcomes and suggestions derive from these case studies but the way forward is also determined by further research needs which will have to be addressed in the future.

16.1 Consumers

The most prominent question is whether consumers' demand for double certified products will increase in future. So far, a growing share of products is certified according to both Fair Trade and organic standards. In the period 2011–2014, around 35% of the global Fair Trade certified production volume was also certified organic. Various reasons apply why consumers opt for these goods, which are subsumed as 'ethical products' and which usually are more expensive. Purchase reasons have been found to be both hedonic/egoistic ones, such as taste, health, GMO-free, as well as altruistic reasons such as social and environmental aspects of production. Nevertheless, both food market segments are interrelated, since they address similar consumer groups. As a result, a double certification can lead to increased prices.

But are consumers willing to pay for higher prices? Although there is price sensitivity and food prices matter to them, the consumers' willingness to pay (WTP) for the organic label has been found to be very high, followed by Fair Trade. But is there a diversion between the estimated and the real WTP? Nevertheless, comparing the WTP for organic versus Fair Trade may also provide some evidence for the degree of consumer confidence or knowledge about what the respective labels purport to deliver. Next to Fair Trade consumers, organic consumers have been observed to be increasingly dissatisfied with anonymous, homogenous food, which may be produced under unknown social conditions. They want greater traceability and information about the diverse origins and conditions under which food is produced and from where and how it is transported. But how will an increasing traceability be perceived by consumers? Will additional information on traceability confuse consumers even more and result in label fatigue?

Communication is critical in establishing credibility of Fair Trade and organic certified products. Information availability regarding the right labels is crucial in ensuring avoidance of product purchases with fake labels. Moreover,

*Corresponding author; e-mail: parvathi@ifgb.uni-hannover.de

consumers need to be made aware that not all Fair Trade products are organic by default, nor do all organic products meet Fair Trade standards. Hence, establishing reliability and trustworthiness through continuous monitoring plays an important role in consumers' WTP for an expensive double certified product.

Furthermore, although Fair Trade and organic markets have shown steady growth in recent decades, these certified products are primarily sold in industrialized countries, but almost 75% of global organic producers reside in the developing continents of Asia, Africa and Latin America. Likewise, there are more Fair Trade cooperatives in these continents. Although this shows the extent of supply and demand interdependence between developing and developed nations, it also highlights the need for the development of domestic Fair Trade and organic markets. Building local niche markets that promote eco-friendly and ethical produce is vital for the future sustainability of these systems.

16.2 Producers

Is small really beautiful? One must be realistic about optimal farm sizes that can be sustained in the future. Contrary to expectations, it is not feasible for all smallholder organic farmers with 2 ha or less to be self-sufficient in meeting their input resource requirements. Their dependence on markets for supply of biopesticides and biofertilizers is inevitable, but organic inputs purchased from the market can be expensive. They may not be within financial budgets of small-scale farmers. Moreover, economies of scale may be difficult to achieve with small farm sizes.

In this scenario, can part-time faming be an option for the needs of smallholder farmers to be deliberated? Given the rural–urban migration in many developing countries and rising agricultural labour costs, smallholder farming may need to evolve from being the only survival strategy to a supplementary source of livelihood. There has to be consistent effort to develop alternative sources of off-farm employment. Schemes have to be introduced to encourage small entrepreneurs. Smallholder farmers need to acquire adequate skills that will enable them to venture out of farming. Training and support on the methods of organic farming, including on composting and crop rotation, need to be given and awareness of Fair Trade standards needs to created even for part-time organic and Fair Trade farmers to make smallholder agriculture viable.

Optimum financial schemes need to evolve to address Fair Trade and organic certification and inspection costs as well as loss of income during the organic conversion period. Effective implementation of contract farming can benefit farmers in developing countries to access domestic and international supermarket chains that normally prefer to order large volumes.

The most pertinent question is whether small-scale farmers benefit from certifying their produce as Fair Trade as well as organic. When smallholders are certified according to organic or Fair Trade standards, they usually produce under contract for large cooperatives or agro-businesses. While mostly positive effects for the farmers have been found for single certifications, these effects do not necessarily persist, because contracts fail and certifications are not renewed. This is also why welfare impacts of Fair Trade and organic systems differ from case to case. Individual farmer behaviour is crucial for the long-term success of farmer–business relationships and hence the success of certifications in improving farmer incomes and ecological sustainability of farming. The organizational capacities of the certifying cooperatives need to be strengthened for certification to make a dent on poverty and vulnerability. Furthermore, more producer-level welfare impact studies are needed to ascertain synergy effects of double certification and its impact on non-certified smallholder farmers.

16.3 Institutional Development

Fair Trade needs to extend its crop portfolio beyond cash crops, flowers, handicrafts, beauty products and gold. It should expand to include indigenous grains, vegetables and fruits. Also, although Fair Trade as a concept is largely assumed to govern ethical trade between developed and developing economies, nevertheless awareness needs to be created among producers to also sell local Fair Trade certified produce in domestic niche markets.

There has to be substantive price benefit from a producer perspective for double certification to be a long-term alternative. In some cases, like double certified black pepper, the producer sells at organic market price or at Fair Trade minimum price, whichever is higher. In effect, the producer receives a price advantage from Fair Trade only if the organic market price falls below the Fair Trade minimum price, the chances of which are remote. The individual farmer and the farmer's Fair Trade cooperative still incur organic and Fair Trade inspection and annual certification costs, but can this change to encourage large organic plantations and organic producer organizations/cooperatives to adopt Fair Trade certification systems as well?

Moreover, certification is an agent of change not only in farming practices, but also in farmer–business relations, because it requires upfront commitment and investment from both sides. Success is largely determined by building trust and aligning expectations of farmers and firms. Longer relationships allow for renewal of the certification. This is in the interest of both sides, due to the high initial certification cost. Organic and Fair Trade standards are short-run agents of change for smallholder farmers. Buyers determine whether they translate into long-run success. Due to their high initial certification and adaptation cost, the beneficial effects of certifications manifest themselves primarily over the longer run. Longer-term farmer–buyer relationships allow for positive investments on both sides. Some agrobusinesses manage these farmer relations much better than others. The three Rs – reliability, reputation and respect – constitute the basis for success.

Furthermore, the integration of different establishments involved in Fair Trade and organic certification systems needs to be addressed to improve value, especially to smallholder farmers. Perhaps ancillary standards can be developed that make it easier for small producer organizations selling organic produce to venture into Fair Trade networks. Similarly, Fair Trade cooperatives may be incentivized to produce under organic agricultural settings. This will encourage farmers to exploit the benefits that these systems have to offer and simultaneously contribute towards sustainable and ethical movements in global agriculture. Although due diligence has to be exercised, nevertheless this concept needs to be at least discussed and debated by the players concerned with Fair Trade and organic agriculture systems.

Index

Note: Page numbers in **bold** type refer to **figures**
Page numbers in *italic* type refer to *tables*
Page numbers followed by 'n' refer to notes